Raspberry Pi Image Processing Programming

With NumPy, SciPy, Matplotlib, and OpenCV

Second Edition

Ashwin Pajankar

Apress®

Raspberry Pi Image Processing Programming

Ashwin Pajankar
Nashik, Maharashtra, India

ISBN-13 (pbk): 978-1-4842-8269-4   ISBN-13 (electronic): 978-1-4842-8270-0
https://doi.org/10.1007/978-1-4842-8270-0

Managing Director, Apress Media LLC: Welmoed Spahr
Acquisitions Editor: Celestin Suresh John
Development Editor: James Markham
Coordinating Editor: Aditee Mirashi
Copyeditor: April Rondeau

Cover designed by eStudioCalamar

Cover image designed by Freepik (www.freepik.com)

Distributed to the book trade worldwide by Springer Science+Business Media New York, 1 New York Plaza, Suite 4600, New York, NY 10004-1562, USA. Phone 1-800-SPRINGER, fax (201) 348-4505, email orders-ny@springer-sbm.com, or visit www.springeronline.com. Apress Media, LLC is a California LLC and the sole member (owner) is Springer Science+Business Media Finance Inc (SSBM Finance Inc). SSBM Finance Inc is a Delaware corporation.

For information on translations, please e-mail booktranslations@springernature.com; for reprint, paperback, or audio rights, please e-mail bookpermissions@springernature.com.

Apress titles may be purchased in bulk for academic, corporate, or promotional use. eBook versions and licenses are also available for most titles. For more information, reference our Print and eBook Bulk Sales web page at http://www.apress.com/bulk-sales.

Any source code or other supplementary material referenced by the author in this book is available to readers on GitHub (github.com/apress) For more detailed information, please visit http://www.apress.com/source-code.

Printed on acid-free paper

This book is dedicated to the memory of
Prof. Govindarajulu Regeti
(July 9, 1945 to March 18, 2021)

Popularly known to everyone as RGR, Prof. Govindarajulu obtained a Bachelor of Technology degree in electrical and electronic engineering from JNTU Kakinada. He also earned a master's degree and Ph.D. from IIT Kanpur. Prof. Govindarajulu was an early faculty member of IIIT Hyderabad and played a significant role in making IIIT Hyderabad the top-class institution that it is today. He was by far the most loved and cheered for faculty member of the institute. He was full of energy to teach and full of old-fashioned charm. There is no doubt he cared for every student as an individual, taking care to know about and to guide them. He has taught, guided, and mentored many batches of students at IIIT Hyderabad (including the author of this book).

Table of Contents

About the Author

Ashwin Pajankar earned a master of technology in computer science engineering from IIIT Hyderabad and has over 25 years of experience in the area of programming. He started his journey in programming and electronics at the tender age of seven with an MS-DOS computer and BASIC programming language. He is now proficient in Assembly programming, C, C++, Java, shell scripting, JavaScript, Go Programming, HTML, and Python. His other technical expertise includes single-board computers such as Raspberry Pi and Banana Pro, microcontroller boards such as Arduino, and embedded boards such as BBC Micro Bit. He has extensively worked on domains such as software/product testing, software automation, databases, data analytics and visualization, computer vision, and web development.

He is currently a freelance online instructor teaching programming and electronics to more than 82,000 professionals. He also regularly conducts live programming bootcamps for software professionals. His growing YouTube channel has an audience of more than 11,000 subscribers. He has published more than 20 books on programming and electronics with many international publishers and is writing more books with Apress. He also regularly reviews books on the topics of programming and electronics written by other authors.

Apart from his work in the area of technology, he is active in the community as a leader and volunteer for many social causes. He has won several awards at his university (IIIT Hyderabad) and also at past workplaces for his leadership in community service for uplifting the underpriviledged with education and skill-based training and employment assistance. He has also participated in many industry institute linkage programs and connected his past employers with his alma maters. During the COVID-19 pandemic (which was continuing at the time of writing of this book), he participated in and led many initiatives to distribute essential supplies such as food, clothes, and medicine to the needy people in his local community. He regularly conducts initiatives for assisting homeless people with clothes, essential supplies, and medical care.

About the Technical Reviewer

 **Lentin Joseph** is an author, roboticist, and robotics entrepreneur from India. He runs a robotics software company called Qbotics Labs in Kochi/Kerala. He has ten years of experience in the robotics domain, primarily in the Robot Operating System, OpenCV, and PCL.

He has authored ten books on ROS, namely, *Learning Robotics Using Python*, first and second editions; *Mastering ROS for Robotics Programming*, first and second editions; *ROS Robotics Projects*, first and second editions; ROS Programming: Building Powerful Robots; and *Robot Operating System (ROS) for Absolute Beginners* first and second editions.

He is also co-editor of the book *Autonomous Driving and Advanced Driver-Assistance Systems (ADAS): Applications, Development, Legal Issues, and Testing*.

He obtained his masters in robotics and automation from India and has also worked at the Robotics Institute, Carnegie Mellon University, USA. He is also a TEDx speaker.

Acknowledgments

I would like to thank my friend Anuradha for encouraging me to share my knowledge with the world. I would like to thank my longtime mentors from Apress, Celestin and Aditee, for giving me an opportunity to share my knowledge and experience with readers. I thank Mark Powers and James Markham for helping me to shape this book as per Apress standards. I am in debt to the technical reviewer, Lentin Joseph. I also thank Prof. Govindrajulu Sir's family, Srinivas (son) and Amy (daughter-in-law), for allowing me to dedicate this book to his memory and sharing with us Govindrajulu Sir's biographical information and his photograph for publication. I would also like to thank all the people associated with Apress who have been instrumental in bringing this project to reality.

Introduction

I want to keep this introduction short, concise, and precise. I have been working with the domain of image processing for quite a while now. I was introduced to Python more than 15 years ago. When I first worked with image processing using Raspberry Pi, I found it a bit tedious to comb through all the literature available as printed books, video tutorials, and online tutorials, as most lacked a step-by-step yet comprehensive guide for a beginner getting started. It was then that I resolved to write a book, and then I published the first edition of this book. It has been almost five years since the first edition was published and it needed a lot of updates.

This book is the fruit of my efforts and cumulative experience of thousands of hours (apart from the ones spent writing the actual book) spent going through technical documentation, watching training videos, writing code with the help of different tools, debugging faulty code snippets, posting questions, and participating in discussions on various question-answer and technical forums online, and referring to various code repositories for directions. I have written this edition of the book in such a way that beginners will find it easy to understand the topics. This book has hundreds of code examples and images (of outputs of code execution and screenshots) so readers can understand each concept perfectly. All the code examples are adequately explained.

The book begins with a general discussion of Raspberry Pi and Python programming. It is followed by a discussion of concepts related to the domain of image processing. Then, it explores the libraries Pillow and TKinter. The following few chapters focus on the scientific Python ecosystem and image processing with libraries such as NumPy, SciPy, and Matplotlib. The last chapter discusses how OpenCV can be used for applying image-processing routines on a live webcam feed. The appendix briefly discusses the pgmagik library, which can be used for generating sample images. It also has plenty of tips and tricks for using a Raspberry Pi board effectively.

As a final note, we started and finished working on this project at a very turbulent time (early 2022) when much of the world was facing yet another wave of fresh infections brought by new variants of the virus causing the COVID pandemic. Everyone around us is still recovering from the impact of the pandemic. This book offered me a sense of purpose. I hope that all the readers enjoy reading this book as much as I enjoyed writing it.

CHAPTER 1

Introduction to Single-Board Computers and Raspberry Pi

Let's start this exciting journey of exploring the scientific domain of digital image processing with Raspberry Pi. To begin, we must be comfortable with the basics of single-board computers (SBCs) and the Raspberry Pi. This chapter discusses the definition, history, and philosophy of SBCs. It compares SBCs to regular computers, then moves toward the most popular and bestselling SBC of all time, the Raspberry Pi. By the end of this chapter, we will have adequate knowledge to independently set up our own Raspberry Pi. This chapter aims to make us comfortable with the basic concepts of SBCs and Raspberry Pi setup.

Single-Board Computers (SBCs)

A single-board computer (referred to as an SBC from now on) is a fully functional computer system built around a single printed circuit board. An

© Ashwin Pajankar 2022
A. Pajankar, *Raspberry Pi Image Processing Programming*,
https://doi.org/10.1007/978-1-4842-8270-0_1

SBC has a microprocessor(s), memory, input/output, and other features required of a minimally functioning computer. Unlike with desktop personal computers (PC), most SBCs do not have expansion slots for peripheral functions or expansion. As all components—processor(s), RAM, GPU, etc.—are integrated on a single printed circuit board (PCB), we cannot upgrade an SBC.

Few SBCs are made to plug into a backplane for system expansion. SBCs come in many varieties, sizes, shapes, form factors, and feature sets. Due to advances in electronics and semiconductor technologies, most SBC prices are very low. With a price of around $50 a piece, we have in our hands a development tool suitable for new applications, hacking, debugging, testing, hardware development, and automation systems.

SBCs are usually manufactured with the following form factors:

- Pico-ITX

- PXI

- Qseven

- VMEbus

- VPX

- VXI

- AdvancedTCA

- CompactPCI

- Embedded Compact Extended (ECX)

- Mini-ITX

- PC/104

- PICMG

Differences Between SBCs and Regular Computers

Table 1-1 lists the differences between SBCs and regular computers.

Table 1-1. *Differences Between SBCs and Regular Computers*

Single-Board Computer	Regular Computer
Not modular	Modular
Components cannot be upgraded or replaced	Components can be upgraded or replaced
A system on chip	Not a system on chip
Has a small form factor	Has a large form factor
Is portable	Is mostly non-portable or semi-portable
Consumes less power	Consumes more power
Cheaper than a regular computer	Costs more than an SBC

System on Chips (SoCs)

SBCs are predominantly systems on chips (SoCs). A system on a chip (SoC) is an integrated circuit (IC) that has all the components of a computer on a single chip. SoCs are very common with mobile electronic devices because of their low power consumption and versatility. SoCs are widely used in mobile phones, SBCs, and embedded hardware. An SoC includes all the hardware and software needed for its operation.

SoC versus Regular CPU

The biggest advantage of using an SoC is its size. If we use a CPU, it's very hard to make a compact computer because of the number of individual chips and other components that we need to arrange on a board. However, when using SoCs, we can place complete application-specific computing systems in smartphones and tablets, and still have plenty of space for batteries, the antenna, and other add-ons required for remote telephony and data communication.

Due to the very high level of integration and the compact size, an SoC uses considerably less power than a regular CPU. This is a significant advantage of SoCs when it comes to mobile and portable systems. Also, reducing the number of chips by eliminating redundant ICs on a computer board results in a compact board size.

History of SBCs

Dyna-Micro was the first true SBC. It was based on the Intel C8080A and used Intel's first EPROM, the C1702A. The Dyna-Micro was rebranded and marketed by E&L Instruments of Derby, Connecticut, in 1976 as the MMD-1 (Mini-Micro Designer 1). It became famous as the leading example of a microcomputer. SBCs were very popular in the earlier days of computing, as many home computers were actually SBCs. However, with the rise of PCs, the popularity of SBCs declined. Since 2010, there has been a resurgence in the popularity of SBCs due to their lower production costs.

Apart from the MMD-1, here are a few other popular historical SBCs:

- The BBC Micro was built around an MOS technology 6502A processor running at 2MHz.

- The Ferguson Big Board II was a Zilog Z80-based computer running at 4MHz.

- The Nascom was another Zilog Z80-based computer.

Popular SBC Families

Based on their manufacturers and designers, SBCs are grouped into families, models, and generations. Here are a few popular SBC families:

- Raspberry Pi by the Raspberry Pi Foundation
- Banana Pi and Banana Pro
- Intel Up Squared Kits
- CubieBoard
- BeagleBone

The Raspberry Pi

The Raspberry Pi is a family of credit card–sized SBCs developed in the United Kingdom by the Raspberry Pi Foundation. The Raspberry Pi Foundation was formed by Eben Upton in 2009. The aim in developing the Raspberry Pi was to promote the teaching of basic computer science in schools and developing countries by providing a low-cost computing platform.

Raspberry Pi Foundation's Raspberry Pi was released in 2012. It was a massive hit and sold over two million units in two years. Subsequently, the Raspberry Pi Foundation released revised versions of the Raspberry Pi. They also released other accessories for the Pi.

More information about the Raspberry Pi Foundation can be found on their website at `https://www.raspberrypi.org`.

The product page for Raspberry Pi's current production models and other accessories can be found at `https://www.raspberrypi.org/products`.

I have written, executed, and tested all the code examples in this book on Raspberry Pi 4 Model B units with 8GB RAM. Table 1-2 lists the specifications of the Raspberry Pi 4 Model B.

Table 1-2. *Specifications of the Raspberry Pi 4 Model B*

Release Date	June 2019
Architecture	ARMv8
SoC broadcom	BCM2711
CPU	Quad core Cortex-A72 (ARM v8) 64-bit SoC @ 1.5GHz
GPU	Broadcom VideoCore IV
Memory	2GB, 4GB or 8GB LPDDR4-3200 SDRAM (depending on model)
USB	2 USB 3.0 ports, 2 USB 2.0 ports
Video output	2 × micro-HDMI ports (up to 4kp60 supported) 2-lane MIPI DSI display port
On-board storage	Micro SDHC slot
On-board network	2.4 GHz and 5.0 GHz IEEE 802.11ac wireless Bluetooth 5.0, BLE Gigabit Ethernet
Power source	3A 5V via MicroUSB

Figure 1-1 shows the front view of the Raspberry Pi 4 Model B.

Figure 1-1. *The front view of Raspberry Pi 4 Model B board[1]*

The image clearly shows all the important components of the board. It shows USB ports, Ethernet port, micro-HDMI ports, 3.5mm audio jack, USB type C port for power, and connectors for CSI and DSI.

Figure 1-2 shows the top view of the board.

---

[1] Image provided by Laserlicht under a CC-by-SA 4.0 license (`https://creativecommons.org/licenses/by-sa/4.0/deed.en`)

Figure 1-2. *The top view of Raspberry Pi 4 Model B board*[2]

We can see the microSD card slot clearly in the bottom view (Figure 1-3).

---

[2] Image provided by Laserlicht under a CC-by-SA 4.0 license (https:// creativecommons.org/licenses/by-sa/4.0/deed.en)

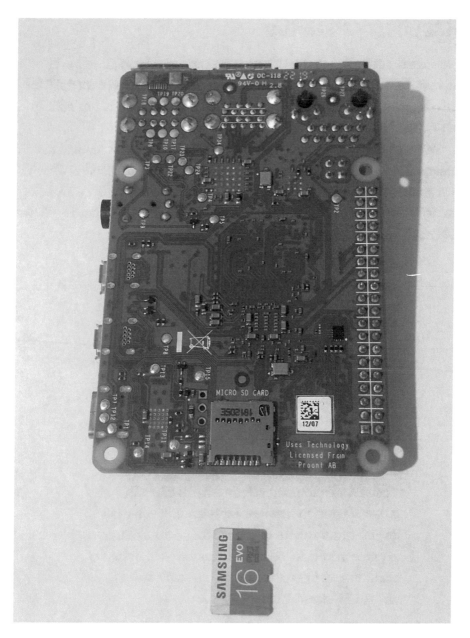

Figure 1-3. *The bottom view of a Raspberry Pi 4 Model B board with a microSD card*

Raspberry Pi Setup

We have to set up Raspberry Pi before we can use it for exploration and adventure. As mentioned earlier, I am using a Raspberry Pi 4 Model B for this setup. Make sure that you use a model of RPi board that has built-in Wi-Fi. The setup process is almost the same for all models.

We are going to use the Raspberry Pi board in headless mode. This means that we will not connect any keyboard, mouse, or display. We will just connect it to the Wi-Fi network and access it remotely. So, we are not going to need a lot of hardware for this. Apart from the RPi board itself, we will need the following things:

> **A computer** – We will need a computer with **a** Windows, macOS, Linux, UNIX, or BSD operating system. We need it for preparing the microSD card for the RPi board.

> **Wi-Fi connection with internet** – We need the internet to install the operating system on the microSD card. We will need it further to connect the RPi to the internet and install necessary software on the RPi.

> **microSD card** – We will need a microSD card. I usually recommend at least a 32GB Class 10 microSD card for the best experience. If you wish to know more about the classes of microSD cards, please visit the following URL on the web: `https://www.sdcard.org/developers/sd-standard-overview/speed-class/`

> **microSD card reader** – We need this so that the computer can read from and write to the microSD card. We need to modify the contents of the

microSD card for headless setup. There are many variations of microSD card reader. The following (Figure 1-4) is an image of a universal card reader.

Figure 1-4. *Universal card reader*[3]

A power supply – We need a power supply to power the RPi board. RPi 4B uses a USB Type C power supply like the one at `https://www.raspberrypi.com/products/type-c-power-supply/`, and a few other models use a micro USB power supply like the one at `https://www.raspberrypi.com/products/micro-usb-power-supply/`.

We can procure all these things and the RPi boards at online marketplaces or local hobby electronics stores.

---

[3] Image provided by Alen under a CC0 1.0 Universal (CC0 1.0) license (`https://creativecommons.org/publicdomain/zero/1.0/`)

Prepare the microSD Card

Following are the steps to prepare the card:

1) Insert the card into the card reader and connect it to the computer.

2) Install the **Raspberry Pi Imager** software from https://www.raspberrypi.com/software/. For Debian and derivatives, run the following command in the terminal emulator:

```
sudo apt install rpi-imager
```

3) Open the installed **Raspberry Pi Imager** program. It looks as shown in Figure 1-5.

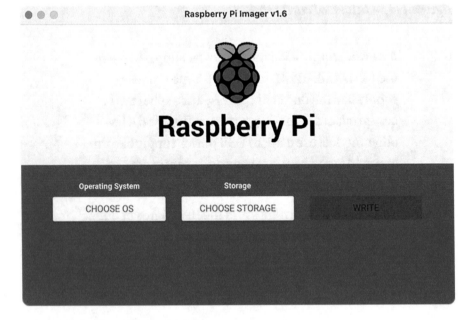

Figure 1-5. *Raspberry Pi Imager*

4) Click on the **CHOOSE OS** button. It opens options as shown in Figure 1-6.

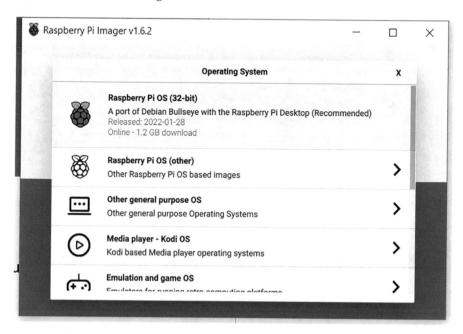

Figure 1-6. *Choosing the OS*

Choose the option **Raspberry Pi OS (other)** by clicking on it. It shows another list of options, as shown in Figure 1-7.

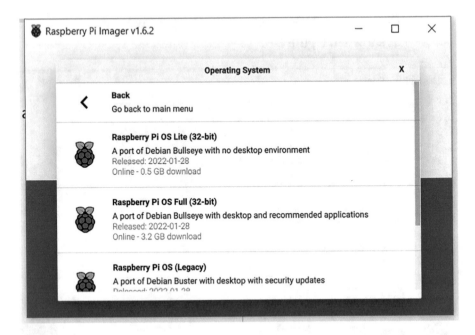

Figure 1-7. *Choosing the OS*

Choose **Raspberry Pi OS (Legacy)**. It is a very well-maintained flavor of **Debian Buster** that supports command-line utilities for the Raspberry Pi camera.

5) Now, click on the **CHOOSE STORAGE** button. It will show the microSD card connected to the computer, as shown in Figure 1-8.

Figure 1-8. *Choosing the storage*

6)   It will enable the **WRITE** button, as shown in
     Figure 1-9.

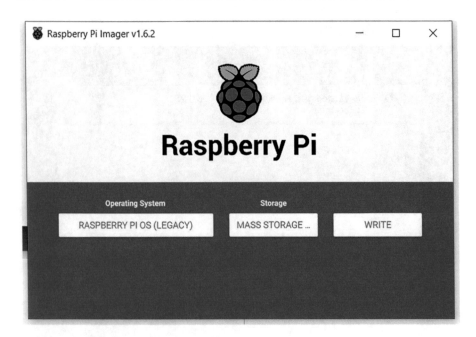

Figure 1-9. *Write button enabled*

Click the **WRITE** button. It will show a warning message, as shown in Figure 1-10.

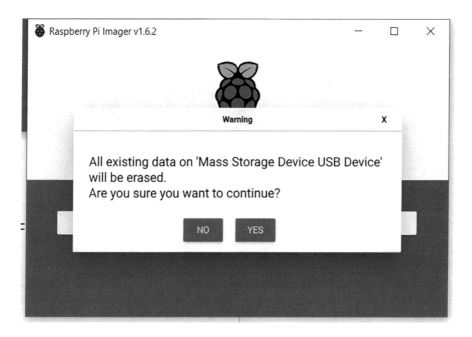

Figure 1-10. *Warning message*

Don't worry about the message. Click the **YES** button.
It starts downloading the files from the internet and
writing them to the microSD card. A progress bar will
appear, as shown in Figure 1-11.

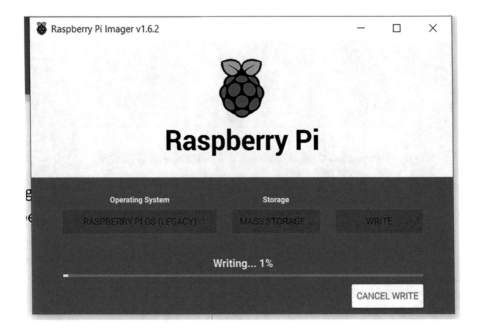

Figure 1-11. *Progress bar*

Once the entire process completes, it shows the
following (Figure 1-12) message.

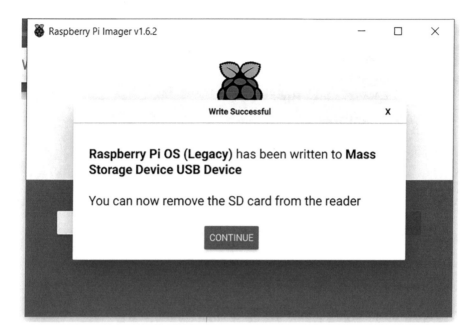

Figure 1-12. *Final message*

Click the **CONTINUE** button. Disconnect the card reader and reconnect it again.

7) Once we reconnect it, it will show in the filesystem as a drive labeled **boot**. The OS may automatically assign a letter to that. Create an empty file in the location **boot** and name it **ssh**. Don't use any extension. This will enable remote login.

8) Create another file and name it **wpa_supplicant. conf**. It is necessary to connect the RPi board to the Wi-Fi after booting up. Add the following contents to the file and save it:

```
country=IN
ctrl_interface=DIR=/var/run/wpa_supplicant
GROUP=netdev
```

```
update_config=1

network={
scan_ssid=1
ssid="TP-Link_710E"
psk="internet1"
}
```

Remember to change the SSID and PSK sections with the name and the password of your Wi-Fi.

9) Now, safely disconnect the microSD card reader.

Booting Up Raspberry Pi

This is the easiest part. Following these steps:

1) Remove the microSD card from the card reader and insert it into the microSD card slot of the RPi board.

2) Make sure that the main power switch is in the OFF position at this point. Connect the RPi to the power supply.

3) Switch on the power switch.

We can see the lights on the RPi board blinking at this point. This means that the RPi is booting up. Wait for a couple of minutes for the boot process to complete.

Accessing Raspberry Pi Remotely

We have booted up the RPi board in headless mode. This means that we have not directly connected any I/O devices such as display, keyboard, or mouse to it. Let's connect to it remotely. First, we must know its IP address.

Since we enabled the SSH and provided it the settings of our Wi-Fi before booting up, it is connected to the Wi-Fi at the time of the booting process. There are many ways to find out the IP address. If you are a part of an organization (workplace, research lab, or university), check with your network/system administrator to find out the IP address of the RPi board. If it is your personal Wi-Fi, then on all the UNIX-like systems (Linux, BSD, macOS), you can run a command to find out the IP address. For Debian and derivatives (Ubuntu, Raspberry Pi OS, etc.), we can install the following command:

```
sudo apt install nmap -y
```

This utility (NMAP) scans the network. The command to find out all the IP addresses connected to the network is as follows:

```
sudo nmap -sn 192.168.0.*
```

Change the numbers to represent your own network. It produces the following output:

```
Starting Nmap 7.80 ( https://nmap.org ) at 2022-02-01 11:26 IST
Nmap scan report for 192.168.0.1
Host is up (0.0025s latency).
Nmap scan report for 192.168.0.100
Host is up (0.0023s latency).
Nmap scan report for 192.168.0.102
Host is up (0.091s latency).
Nmap done: 256 IP addresses (3 hosts up) scanned in
2.72 seconds
```

You can find the IP address of your own system with the commands ipconfig (Windows) or ifconfig (UNIX-like systems). From this list, we can eliminate all the known devices with known IP addresses. If there are too many devices attached to the home network, then turn off the Wi-Fi of all the unnecessary devices. In my case, the IP address of the RPi board is 192.168.0.100.

On a Windows computer, we can install the Zenmap utility, which is the graphical interface for the **nmap**. We can download and install it from `https://nmap.org/zenmap/`. It is also available for UNIX-like systems. Install the utility and open it. It looks as shown in Figure 1-13.

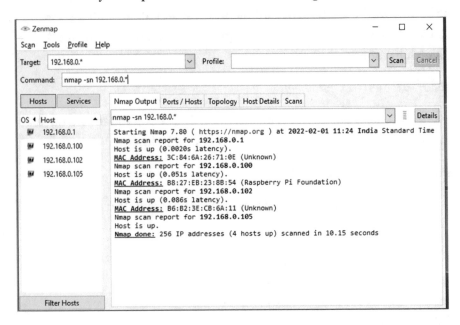

Figure 1-13. *Zenmap GUI*

We can easily use it. Mention 192.168.0.* in the **Target** text box. Mention the command that we used earlier (nmap -sn 192.168.0.*) in the **Command** text box. And then click the **Scan** button. It takes some time for the scan to finish. Sometimes the scan fails prematurely. In such cases, keep trying, and it will work after a couple of attempts. It is all open source and free, so we cannot complain. It produces the result shown in Figure 1-13 once finished. Sometimes, it also shows the names of the manufacturers of the connected devices beside their MAC addresses. This makes it easy to identify the RPi board.

Once we have the IP address of the RPi board, we can log in remotely. We can use a variety of SSH tools, such as the built-in ssh utility on UNIX-like operating systems, to remotely log in to the RPi board. Just run the following command in the terminal emulator of your operating system:

```
ssh pi@192.168.0.100
```

The **pi** is the username. It will prompt for the password. It is **raspberry**. This way, we can remotely log in to the RPi board and run commands.

PuTTY and **Bitvise SSH** clients are a few popular ones on Windows. However, I find **MobaXterm** very handy for this. It has the options for remote terminal, file transfer, and X-11 forwarding. We do not have to install anything else if we have this tool. Install it from https:// mobaxterm.mobatek.net/. Once installed, open it. The window looks as shown in Figure 1-14.

Figure 1-14. *MobaXterm*

In this window, click on the **Start local terminal** button. It launches a local terminal window. There, we can run the following command to connect to the RPi board:

```
ssh pi@192.168.0.100
```

The **pi** is the username. It will prompt for the password. It is **raspberry**. This way, we can remotely log in to the RPi board and run commands.

Now, it is time to brush up on the UNIX concepts and commands that we learned in school. We can run all the Linux/UNIX commands on this remote terminal. One of the best things about the **MobaXterm** is it has a built-in **XServer** program that is automatically launched when we launch this utility. We can see the symbol for that in the top-right corner of the window. UNIX-like systems allow us to access the GUI applications remotely with X-11 forwarding. In RPi, it is enabled by default. We just need an **XServer** utility at the other end (the computer from which we are connecting to the RPi, or any other UNIX-like system). As discussed already, we have it with the **MobaXterm**. Just run the following command in the terminal:

```
nohup pcmanfm &
```

It opens a window for the **File Explorer** utility, as shown in Figure 1-15.

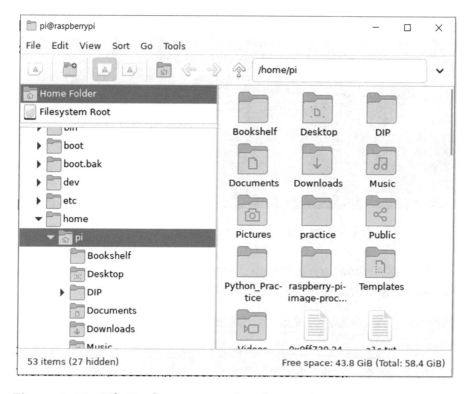

Figure 1-15. *File Explorer accessed with X-11 forwarding*

Keep in mind that this **File Explorer** utility is running on the RPi. We are accessing its GUI using X-11 Forwarding. This way, with the combination of nohup (no hangup) and &, we can launch any GUI utility from the terminal. & makes sure that the control of the terminal is returned back to us, and nohup keeps the process running even when the user logs out.

Configuring Raspberry Pi

Now, type sudo raspi-config in the prompt and press Enter. It opens the tool **raspi-config**, which is the configuration tool for the Raspberry Pi OS. First, update it, as shown in Figure 1-16.

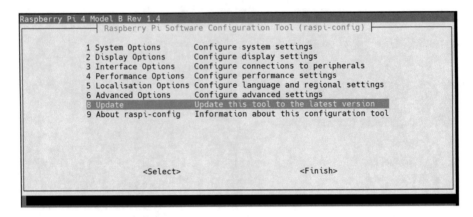

Figure 1-16. *Updating the raspi-config tool*

It will take some time to update. Once updated, go to the fifth option for Localization. It looks as follows (Figure 1-17).

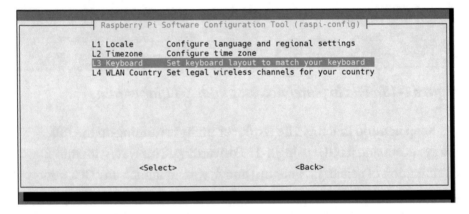

Figure 1-17. *Localization*

Set all these options as per your choice, and in the main menu choose **Finish**. It will ask to reboot. Choose **Yes**, and it will reboot the RPi.

The Raspberry Pi OS

An operating system is the set of basic programs and utilities that make a computer work. It is an interface between the user and the computer. Raspberry Pi OS is a free operating system based on the popular Linux distribution Debian. It is optimized for the Raspberry Pi family of SBCs. It is even ported to the other, similar SBCs like Banana Pro.

The config.txt File

Raspberry Pi does not have a conventional BIOS or UEFI. The BIOS (basic input/output system) is the program that a computer's microprocessor uses to get the computer system started after it is turned on. It also manages data flow between the computer's operating system and attached peripheral devices, such as the hard disk, video adapter, keyboard, mouse, and printer.

Since the Raspberry Pi does not have a BIOS/UEFI, the various system configuration parameters that are normally stored and modified using the BIOS/UEFI are instead stored in a text file called **config.txt**.

The Raspberry Pi config.txt file is on the boot partition of the Raspberry Pi. It is normally accessible as **/boot/config.txt** from Linux. However, from Windows and MacOS, it is seen as a file in the accessible part of the microSD card. The accessible part of the card, as we already know, is labeled **boot**.

On a Raspberry Pi, you can edit this file with the following command:

```
sudo nano /boot/config.txt
```

> **Note**   **nano** is a simple and easy-to-learn terminal-based text editor for Linux. Visit its home page at `https://www.nano-editor.org` to learn more about it. I find it easier to use than **vi** or **vim** editors.
>
> To learn more about **config.txt**, visit the page `http://elinux.org/RPiconfig`. A sample configuration can also be found at `http://elinux.org/R-Pi_configuration_file`.

Updating the RPi

The Raspberry Pi must be connected to the internet in order to update it successfully. Let's update the firmware and the Raspberry Pi OS.

Updating the Firmware

To update the firmware, run the command `sudo rpi-update`.

Updating and Upgrading Raspberry Pi OS

We will use **APT** for this. **APT** (Advanced Package Tool) is a program that handles the installation and removal of software on Debian and other Debian derivatives. **APT** simplifies the process of managing software on Debian systems by automating the fetching, configuration, and installation of software packages. We need an internet connection for this too.

First, update the system's package list by entering the following command in the terminal:

```
sudo apt update
```

This downloads the package lists from the respective remote repositories and updates them in the local computer so that information on the newest versions of packages and their dependencies is available for the installation and update. It should be run before running the install or upgrade commands.

Next, upgrade all the installed packages to their latest versions using this command:

```
sudo apt full-upgrade -y
```

This fetches new versions of the packages on the local machine that are marked for upgrade. It also detects and installs any dependencies. It also removes obsolete packages.

Doing this regularly will keep the Raspberry Pi OS up to date. After entering these commands, it will take a while to update the OS, because these commands fetch the data and the packages from remote repositories on the internet.

Note   The command apt help will list all the options associated with the APT utility.

Shutting Down and Restarting RPi

We can shut down the RPi safely using either of the following commands:

```
sudo shutdown -h
sudo init 0
```

We can reboot the RPi using either of the following commands:

```
sudo reboot -h
sudo init 6
```

Summary

In this chapter, we learned how to set up and access RPi in headless mode over Wi-Fi. We are comfortable with this part now.

The next chapter will focus on the concepts involved with digital **image processing**. We will also write basic programs with Python 3.

CHAPTER 2

Introduction to Python and Digital Image Processing

In the last chapter, we explored the amazing world of single-board computers and Raspberry Pi. We booted up the Raspberry Pi, connected it to the internet, and updated the Raspberry Pi OS.

In this chapter, we will get started with Python and the concepts of **digital image processing** (DIP). Let's begin with an introduction to Python. I personally find Python amazing and have been enchanted by it. Python is a simple yet powerful programming language. When programmers use Python, it's easy to focus on solving a given problem as they do not have to worry about the syntax. Python perfectly fits the philosophy of Raspberry Pi, which is programming for everyone. That's why it's the most preferred programming platform for Raspberry Pi and other computers.

The following is a list of topics we will learn in this chapter:

- History of Python

- Features of Python

- Python 3

© Ashwin Pajankar 2022
A. Pajankar, *Raspberry Pi Image Processing Programming*,
https://doi.org/10.1007/978-1-4842-8270-0_2

- IDEs for Python

- Introduction to digital image processing

By the end of this chapter, we will be comfortable with the concepts of digital image processing and Python 3 basics.

A Brief History of the Python Programming Language

Python was designed and conceived in the late 1980s. Its actual implementation was started in late 1989 by **Guido van Rossum** at **Centrum Wiskunde & Informatica** (**National Research Institute for Mathematics and Computer Science**) in the Netherlands. Python is a successor to the **ABC programming language**, which itself was inspired by the **SETL programming language**. In February of 1991, Van Rossum publicly published the Python source code to the newsgroup **alt.sources**. The name Python was inspired by the British television show *Monty Python's Flying Circus* because Van Rossum was a big fan of Monty Python.

Van Rossum is the principal author of the Python programming language. He played a central role in guiding the development, enhancement, and further evolution of Python. He held the title Benevolent Dictator for Life for Python till July 2018. Currently, Python development is guided by a five-member **steering council**.

The central philosophy of the Python programming language (**the Zen of Python**) is explained in **PEP-20** (**PEP** stands for **Python Enhancement Proposal**), which can be found at `https://www.python.org/dev/peps/pep-0020`.

It is a collection of 20 software principles, out of which 19 have been documented. The principles are as follows:

- Beautiful is better than ugly.

- Explicit is better than implicit.

- Simple is better than complex.

- Complex is better than complicated.

- Flat is better than nested.

- Sparse is better than dense.

- Readability counts.

- Special cases aren't special enough to break the rules.

- Practicality beats purity.

- Errors should never pass silently.

- Unless explicitly silenced.

- In the face of ambiguity, refuse the temptation to guess.

- There should be one—and preferably only one—obvious way to do it.

- Although that way may not be obvious at first unless you're Dutch.

- Now is better than never.

- Although never is often better than right now.

- If the implementation is hard to explain, it's a bad idea.

- If the implementation is easy to explain, it may be a good idea.

- Namespaces are one honking great idea—let's do more of those!

Features of Python

The following are the features of Python that have made it popular and beloved in the programming community.

Simple

Python is a simple language with a minimalist approach. Reading a well-written and good Python program makes you think you are reading English text.

Easy to Learn

Due to its simple and English-like syntax, Python is extremely easy to learn. That is the prime reason why it is taught as the first programming language to high school and university students who take introductory programming courses. An entire new generation of programmers is learning Python as their first programming language.

Easy to Read

Unlike other high-level programming languages, Python does not obfuscate the code and make it unreadable. The English-like structure of the Python code makes it easier to read compared to code written in other programming languages. This makes it easier to understand and easier to learn compared to other high-level languages like C and C++.

Easy to Maintain

As Python code is easy to read, easy to understand, and easy to learn, anyone maintaining the code becomes comfortable with the codebase very quickly. I can vouch for this from my personal experience maintaining and enhancing large legacy codebases that were written in a combination of Bash and Python 2.

Open Source

Python is an open source project, which means its source code is freely available. You can make changes to it to suit your needs and use the original and modified code in your applications.

High-Level Language

While writing Python programs, you do not have to manage the low-level details like memory management, CPU timings, and scheduling processes. All these tasks are managed by the Python interpreter. You can write the code directly in easy-to-understand English-like syntax.

Portable

The Python interpreter has been ported to many OS platforms. Python code is also portable. All the Python programs will work on the supported platform without requiring many changes if you are careful to avoid system-dependent coding.

You can use Python on GNU/Linux, Windows, Android, FreeBSD, Mac OS, iOS, Solaris, OS/2, Amiga, Palm OS, QNX, VMS, AROS, AS/400, BeOS, OS/390, z/OS, Psion, Acorn, PlayStation, Sharp Zaurus, RISC OS, VxWorks, Windows CE, and PocketPC.

Interpreted

Python is an interpreted language. Let's take a look at what that means. Programs written in high-level programming languages like C, C++, and Java are compiled first. This means that they are first converted into an intermediate format. When we run the program, this intermediate format is loaded from secondary storage (i.e., from the hard disk) to the memory (RAM) by the linker/loader.

So, C, C++, and Java have a separate compiler and linker/loader. This is not the case with Python. Python runs the program directly from the source code. You do not have to bother compiling and linking to the proper libraries. This makes Python programs truly portable, as you can copy the program to one computer from another and the program runs fine as long as the necessary libraries are installed on the target computer.

Object Oriented

Python supports procedure-oriented programming as well as object-oriented programming paradigms.

All the object-oriented programming paradigms are implemented in Python. In object-oriented programming languages, the program is built around objects that combine data and related functionality. Python is a very simple but powerful object-oriented programming language.

Extensible

One of the features of Python is that you can call C and C++ routines from the Python programs. If you want the core functionality of the application to run faster, you can code that part in C/C++ and call it in the Python program (C/C++ programs generally run faster than Python ones).

Extensive Libraries

Python has an extensive standard library that comes pre-installed. The standard library has all the essential features for a modern-day programming language. It has provisions for databases, unit testing (we will explore this later in this book), regular expressions, multi-threading, network programming, computer graphics, image processing, GUI, and other utilities. This is part of Python's batteries-included philosophy.

Apart from the standard library, Python has numerous and evergrowing sets of third-party libraries. The list of these libraries can be found on the Python Package Index.

Robust

Python provides robustness by means of the ability to handle errors. The full stack trace of the encountered errors is available and makes the life of the programmer more bearable. The runtime errors are known as exceptions. The feature that handles these errors is known as an exception-handling mechanism.

Rapid Prototyping

Python is used as a rapid prototyping tool. As you learned earlier, Python has extensive libraries and is easy to learn, which has led many software architects to use it as a tool to rapidly prototype their ideas into working models.

Memory Management

In assembly language and in programming languages like C and C++, memory management is the responsibility of the programmer. This is in addition to the task at hand. This creates an unnecessary burden on the programmer. In Python, the Python interpreter takes care of the memory management. This helps the programmers steer clear of memory issues and focus on the task at hand.

Powerful

Python has everything in it that a modern programming language needs. It is used in applications such as computer visioning, supercomputing, drug discovery, scientific computing, simulation, and bioinformatics. Millions of programmers around the world use Python. Many big organizations like NASA, Google, SpaceX, and Cisco use Python for their applications and infrastructure.

Community Support

I find this to be the most appealing feature of Python. Recall that Python is open source. It also has a community of almost a million programmers throughout the world (probably more, as today high school kids are learning Python too). There are also plenty of forums on the internet to support programmers who encounter a roadblock. None of my queries related to Python have ever gone unanswered.

Python 3

Python 3 was released in 2008. The Python development team decided to do away with some of the redundant features of Python, simplify some more features, rectify some design flaws, and add a few much-needed features.

It was decided that a major revision number was needed for this and that the resultant release would not be backward compatible. Python 2.x and 3.x were supposed to coexist in parallel so the programmer community would have enough time to migrate their code and third-party libraries from 2.x to 3.x. Python 2.x code cannot be run as-is in most cases, as there are significant differences between 2.x and 3.x.

Python 2 and Python 3 on Raspberry Pi OS

Raspberry Pi OS is a derivative of Debian Linux. Python 2 and Python 3 interpreters are pre-installed in Raspberry Pi OS. The Python 2 interpreter can be invoked by running the command python in the terminal. The Python 3 interpreter can be invoked by running the command python3 in the terminal. We can check the Python 3 interpreter version by running either of the following commands:

```
python3 -V
python --version
```

You can check the location of the Python 3 binary by running the following command:

```
which python3
```

Running a Python Program and Python Modes

You have set up your environment for Python programming, so let's get started with the simple concepts of Python. Python has two modes— interactive mode and script mode. Let's look at these modes in detail.

Interactive Mode

Python's interactive mode is a command-line shell. It provides immediate output for every executed statement. It also stores the output of previously executed statements in the active memory. As new statements are executed by the Python interpreter, the entire sequence of previously executed statements is considered while evaluating the current output. We have to type the command python3 in the terminal to invoke the Python 3 interpreter into interactive mode, as follows:

```
pi@raspberrypi:~ $
Python 3.4.2 (default, Oct 19 2014, 13:31:11)
[GCC 4.9.1] on linux
Type "help", "copyright", "credits" or "license" for more
information.
>>>
```

You can execute Python statements directly in this interactive mode just like when you run commands from the OS shell/console, as follows:

```
>>>print ('Hello World!')
Hello World!
>>>
```

We will not be using interactive mode in this book. However, keep it in mind, as it's the quickest way to check small snippets of code (five to ten lines). We can quit interactive mode using the exit() statement, as follows:

```
>>> exit()
pi@raspberrypi:~ $
```

Script Mode

Script mode is where the Python script files (.py) are executed by the Python interpreter.

Create a file called test.py and add the `print ('Hello World!')` statement to the file. Save the file and run it with the Python 3 interpreter as follows:

```
pi@raspberrypi:~ $ python3 test.py
Hello World!
pi@raspberrypi:~ $
```

In this example, `python3` is the interpreter and `test.py` is the filename. In case the Python `test.py` file is not in the same directory where we are invoking the `python3` interpreter, we have to provide the absolute path of the Python file.

IDEs for Python

An integrated development environment (IDE) is a software suite that has all the basic tool s needed to write and test programs. A typical IDE has a compiler, a debugger, a code editor, and a build automation tool. Most programming languages have various IDEs to make programmers' lives better. Python too has many IDEs. Let's look at a few of them.

IDLE

IDLE stands for Integrated DeveLopment Environment. We have to install it on Raspberry Pi OS. IDLE3 is for Python 3. It's popular with Python beginners. We can install it with the following command:

```
pip3 install idle
```

We can launch IDLE3 using the following command:

```
nohup idle &
```

Figure 2-1 shows the IDLE3 code editor and an interactive prompt.

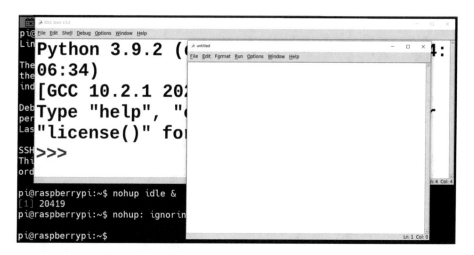

Figure 2-1. *IDLE3 interactive prompt for Python 3 interpreter and a new code file*

We can go to **Options ➤ Configure IDLE** in the main menu. We can change the settings of the editor here (Figure 2-2).

Figure 2-2. *IDLE3 configuration options*

Geany

Geany is a text editor that uses the GTK+ toolkit and has the basic features of an integrated development environment. It supports many file types and programming languages. It has some nice features. Check out `https://www.geany.org` for more details. Figure 2-3 shows the Geany text editor window.

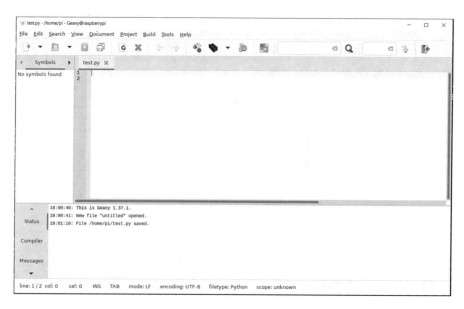

Figure 2-3. *Geany IDE*

Type `print("Hello World!")` in the code editor and save the file in the /**home**/**pi** directory as **test.py**. Click **Build** in the menu bar and then choose **Execute**. You can also use the **F5** keyboard shortcut to execute the program. The program will execute in a terminal window. You have to press **Enter** to close the execution window. The default Python interpreter for Geany is Python 2. We will need to change it to Python 3. To do that, go to **Build ➤ Set Build Commands**. The window shown in Figure 2-4 will appear.

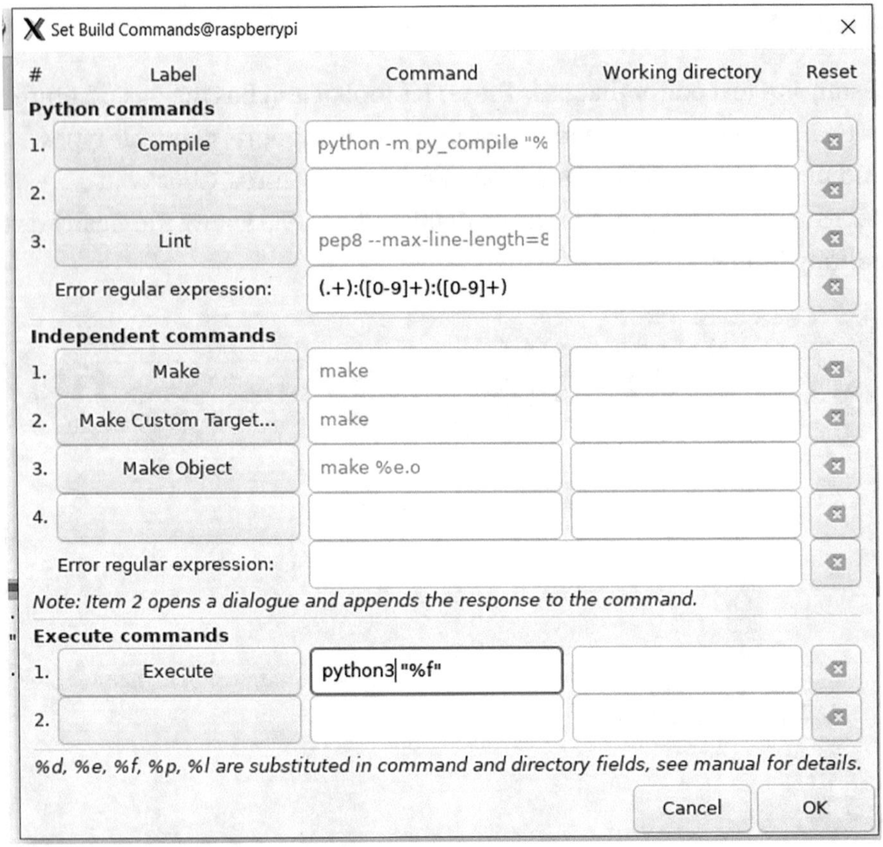

Figure 2-4. Set Build Commands window

In the Set Build Commands window, within the Execute Commands section, change **python "%f"** (highlighted in the box in Figure 2-4) to **python3"%f"** to set the Python 3 interpreter as the default interpreter. After that, run the program again to verify that everything works correctly.

Thonny IDE

Thonny IDE also comes installed with the latest edition of the Raspberry Pi OS. It works with Python 3 by default (and only with Python 3, so we do not have to change the settings). We can invoke it with the following command:

```
nohup thonny &
```

Figure 2-5 shows a screenshot of the Thonny IDE instance.

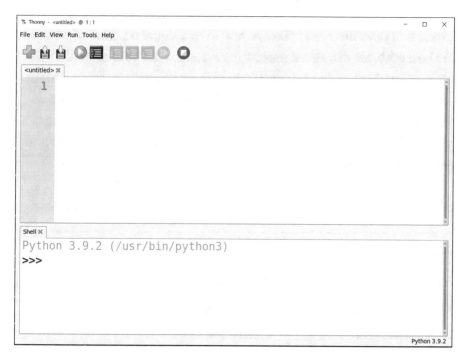

Figure 2-5. *Thonny IDE*

Introduction to Digital Image Processing

To get started with digital image processing (DIP), we will briefly review our understanding of a few basic concepts about related topics.

Signal Processing

Anything that carries any information, when represented mathematically, is called a signal. The process or technique used to extract useful and relevant information from a given signal is known as signal processing. The system that does this type of task is known as a signal processing system. The best example of a signal processing system is the human brain. It processes various types of signals from various senses. The human brain is a biological signal processing system. When the system is made up of electronic components, it is known as an electronic signal processing system. Signal processing is a discipline that combines mathematics and electrical engineering.

There are two types of electronic signals—analog and digital. The following table lists the differences between these two types of signals.

Analog	Digital
A continuous signal	Discrete in nature
Denoted by sinusoidal waves	Denoted by square waves
A continuous signal	A discrete signal
Deteriorated due to noise	Not affected by noise
Not flexible	Flexible and can perform a variety of operations
Consumes less bandwidth	Consumes a lot of bandwidth
Draws a lot of power	Draws less power compared to analog

Analog	Digital
There are lots of errors in capturing, storing, and transmitting analog signals	Fewer errors compared to analog signals, and there are checksum and error correction algorithms available for digital signals.

Image Processing

An image is a signal. So, image processing is a type of signal processing. Image processing systems are types of signal processing systems. The combination of the eye and brain is an example of a biological image processing system. There are two types of image processing systems—analog and digital.

Analog Image Processing

The days of still and motion film cameras represent the analog era. The sources of analog images are film cameras (still and motion), older fax machines, and telex machines. Older television systems, CRT monitors, and film projectors represent analog image processing systems. Analog image processing involves extensively using analog electronic systems, mechanical parts, optics, and chemistry (to develop and store films).

Digital Image Processing

With the advent of digital computers, storage systems, image sensors, and digital cameras, images can be captured, stored, and processed in digital formats. Using digital computers to process and retrieve information from an image (analog and digital images both) is known as digital image processing. It involves extensive use of digital sensors (digital cameras), digital computers, and digital storage devices. Here are some applications of digital image processing systems:

- Computerized image editing, correction, enhancement, denoising, etc.

- Medical image processing and diagnostics assistance

- Space image processing (processing images from Hubble and ground-based telescopes)

- Industrial applications like product inspection and sorting

- Biometrics (finger, face, and iris recognition)

- Filmmaking and visual effects

- Remote sensing (processing images from aerial and satellite sources)

The following scientific disciplines significantly overlap with digital image processing:

- Signal processing

- Digital electronics

- Computer/machine vision

- Biological vision

- Artificial intelligence, pattern recognition, and machine learning

- Robotics and robot vision

Using Raspberry Pi and Python for Digital Image Processing (DIP)

Since DIP requires digital computers, we need to use a computer and an associated programming platform to implement digital image processing. Raspberry Pi fulfills all the requirements of a minimal power system required for DIP. The most appealing feature of Raspberry Pi is its low cost. Also, it can be interfaced with a variety of digital image sensors like webcams and the Pi camera module.

We learned that Python is an easy-to-learn programming language that helps us focus on the task at hand rather than having to worry about syntax. It is also the most preferred programming platform for Raspberry Pi. Many third-party image processing and visualization libraries are available for Python. This makes Raspberry Pi and Python the most obvious choices for beginners to get started with DIP.

Exercise

Complete the following Exercise to better understand the Python 3 background.

- Visit and explore the Python home page at `www.python.org`.

- Visit and explore the Python documentation page at `https://docs.python.org/3/`.

- Check for new features of the latest releases of Python at `https://docs.python.org/3/whatsnew/index.html`.

Summary

In this chapter, we learned about the background, history, and features of Python. We also studied the important differences between Python 2.x and Python 3.x. We learned to use Python 3 in scripting and interpreter modes. We looked at a few popular IDEs for Python and configured Geany for Python 3 on the Pi. Finally, we learned about the field of digital image processing. In the next chapter, we will get started with a popular digital image processing library in Python called Pillow. We will also learn how to use the **Tkinter** library to show images.

CHAPTER 3

Getting Started

In the previous chapter, we learned about the philosophy of Python. We also learned why we should use Python 3, as well as several concepts related to digital image processing. This chapter will look at image processing programming using Python 3 on the Raspberry Pi. We will learn how to connect a Raspberry Pi to various camera sensors in order to acquire images. We will be introduced to the Pillow and Tkinter libraries in Python 3. The following is a list of topics we will learn in this chapter:

- Image sources

- Using a webcam

- The Pi camera module

- Python 3 for digital image processing

By the end of this chapter, we will be comfortable using the cameras, as well as the libraries **Pillow** and **Tkinter**.

Image Sources

To learn how to do digital image processing, we are going to need digital images. There are standard image datasets that are used all around the world; we will use a couple of free-to-use images from the following database: `http://sipi.usc.edu/database/`.

© Ashwin Pajankar 2022
A. Pajankar, *Raspberry Pi Image Processing Programming*,
https://doi.org/10.1007/978-1-4842-8270-0_3

We can read copyright-related information for these images at
`https://sipi.usc.edu/database/copyright.php`.

We are going to use image files **4.1.07.tiff**, **4.1.08.tiff**, **gray21.512**, and
ruler.512 from this database throughout the book for the demonstrations.

In this chapter, we will start programming. We need to organize all the
files in a single directory, with chapter subdirectories to save the code. Run
the following commands in terminal to create a directory structure for the
book code and image datasets:

```
mkdir DIP
cd DIP
mkdir code
mkdir dataset
cd code
mkdir chapter03
```

Extract all the images into the Dataset directory. Now the directory
structure will look like Figure 3-1.

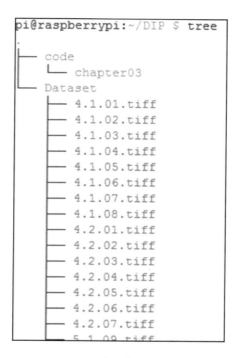

Figure 3-1. *Directory structure for the organization of the dataset and the code*

We use the `tree` command in the DIP directory to view the directory structure.

Note   Please install the `tree` utility with the following command if not installed already:

```
sudo apt install tree
```

The images we just downloaded and extracted are used as standard images for all digital image processing and computer vision research around the world.

Using a Webcam

Let's see how to capture images using a standard USB webcam. The Raspberry Pi 3 has four USB ports. You can use one of these to connect a webcam to the Raspberry Pi. I am using a Logitech c922 USB webcam (see Figure 3-2).

Figure 3-2.   Logitech c922[1]

Note   Before purchasing a webcam, check the webcam's compatibility with the Pi at `http://elinux.org/RPi_USB_Webcams`.

---

[1] Image provided by Pmwiki1 under a CC0 1.0 Universal (CC0 1.0) license (`https://creativecommons.org/publicdomain/zero/1.0/`)

Attach the webcam and run the command lsusb in the terminal. This displays a list of all USB devices connected to the computer. The output will be similar to the following:

```
pi@raspberrypi:~$ lsusb
Bus 002 Device 001: ID 1d6b:0003 Linux Foundation 3.0 root hub
Bus 001 Device 003: ID 046d:085c Logitech, Inc. C922 Pro
Stream Webcam
Bus 001 Device 002: ID 2109:3431 VIA Labs, Inc. Hub
Bus 001 Device 001: ID 1d6b:0002 Linux Foundation 2.0 root hub
```

guvcview

If you can see the USB webcam in the output, it means that the webcam has been detected. There are quite a few Linux utilities with which to capture images using a webcam. If you like GUI, **guvcview** is one of them. Use the following command to install it if it is not already installed:

```
sudo apt install guvcview
```

Once **guvcview** is installed, it can be accessed by running the following command:

```
nohup guvcview &
```

This shows a preview window and the application window, as seen in Figure 3-3.

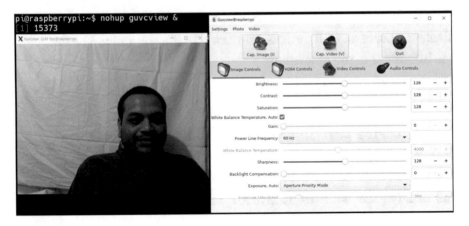

Figure 3-3. *The guvcview utility and camera preview window*

As seen in Figure 3-3, it is a versatile tool that provides a unified interface with which to adjust the properties of the webcam connected to the RPi board. Images and videos can be captured with this tool. As an exercise, please explore this tool further.

fswebcam

The other useful utility is **fswebcam**. It's a command-line utility. Install it using the following command:

```
sudo apt install fswebcam
```

You can invoke the **fswebcam** utility from the terminal to capture an image with the webcam as follows:

```
fswebcam -r 1920x1080 --no-banner test.jpg
```

This will capture an image with the resolution of **1920 x 1080** pixels. The --no-banner flag disables the timestamp banner from the image. The image is saved as **test.jpg**. The output of the command is as follows:

```
pi@raspberrypi:~$ fswebcam -r 1920x1080 --no-banner test.jpg
--- Opening /dev/video0...
Trying source module v4l2...
/dev/video0 opened.
No input was specified, using the first.
--- Capturing frame...
Captured frame in 0.00 seconds.
--- Processing captured image...
Disabling banner.
Writing JPEG image to 'test.jpg'.
```

Then, check in the current directory for the image file called test.jpg.

The Pi Camera Module

The Raspberry Pi Foundation has designed dedicated camera modules for the Raspberry Pi. There are two versions, both of which come in normal and **NoIR** (**No Infrared**) varieties. The original versions are 5-megapixel camera modules. The new versions are 8-megapixel camera modules. Figure 3-4 shows an image of the camera module V2.

Figure 3-4. *The Pi camera module V2[2]*

This one is the normal camera module suitable for normal lighting conditions.

Figure 3-5 shows the NoIR version, suitable for low lighting conditions. More details about them can be found at https://www.raspberrypi.org/ products/camera-module-v2. The camera can be connected to any model of Raspberry Pi by a dedicated camera port, as shown in Figure 3-5.

---

[2] Image provided by the Raspberry Pi Foundation under a CC-by-SA 4.0 license (https://creativecommons.org/licenses/by-sa/4.0/deed.en)

Figure 3-5. *Attaching the Pi camera module to the Pi[3]*

In Figure 3-5, the NoIR camera module V2 is connected to a Raspberry Pi Zero W board (which is a compact board model in the Raspberry Pi family). The Pi camera module connects directly to the GPU. As it is attached to the GPU, there is only a small impact on the CPU, leaving it available for other processing. The difference between the Pi camera and USB webcam is that the Pi camera module has a higher performance level and higher frame rate, with h.264 video encoding.

The Raspberry Pi Foundation has produced another high-quality camera module with interchangeable C- and CS- mounts for lenses. The

---

[3] Image provided by the Raspberry Pi Foundation under a CC-by-SA 4.0 license (https://creativecommons.org/licenses/by-sa/4.0/deed.en)

package includes the camera module and C-to-CS mount adapter. The camera looks as shown in Figure 3-6.

Figure 3-6.   *High-quality camera module[4]*

We can read more about the camera module on the product page at https://www.raspberrypi.com/products/raspberry-pi-high-quality-camera/.

---

[4] Image provided by the Raspberry Pi Foundation under a CC-by-SA 4.0 license (https://creativecommons.org/licenses/by-sa/4.0/deed.en)

raspistill

The command-line utility we use to capture images using the camera module is **raspistill**. We call it as follows:

```
raspistill -o test.png
```

raspistill does not write anything to the console or terminal, so we need to check the directory to see if **test.png** has been created or updated.

This is how you capture images with various cameras and command-line utilities for use in exploring the world of DIP with Pi.

Python 3 for Digital Image Processing

Python 3 does not have any pre-installed libraries for image processing. The **Python Imaging Library** (**PIL**) is one of the most popular beginner-friendly image processing libraries used for image processing in Python.

This library provides extensive file-format support and reasonably powerful image processing capabilities. The **core Image library** is designed for achieving faster access to data stored in a few basic image formats. The drawback with PIL is that after its last stable release (1.1.7) in 2009 and the last commit in 2011, there has not been a new release. Also, PIL does not support Python 3. Officially, we have not heard that PIL is dead. However, there is a friendly fork (derived) project that's an unofficial replacement and enhancement for PIL. It works well with Python 3 and is called **Pillow**. It is mostly backward compatible with PIL. You can get more information about Pillow on its home page at https://python-pillow.org/.

Let's get started with the basics of Pillow. We can install Pillow by running the following command at the terminal:

```
sudo pip3 install pillow
```

This will install the Pillow library for Python 3. To check if it is installed properly, open Python 3 in interpreter mode and run the following sequence of commands. If Pillow is installed properly, these commands will display its version number.

```
Python 3.9.2 (default, Mar 12 2021, 04:06:34)
[GCC 10.2.1 20210110] on linux
Type "help", "copyright", "credits" or "license" for more
information.
>>> from PIL import Image
>>> print(Image.__version__)
8.1.2
>>>
```

Working with Images

Let's work with images now. Before we begin, we will have to install an image viewing utility called **xv** in order for the built-in function show() to work. The problem is that the xv utility is deprecated. So, you will use another utility called xli and point it to xv with a Linux command. Run the following commands:

```
sudo apt-get install xli -y
cd /usr/local/bin
sudo ln -s /usr/bin/xli xv
```

Save the program shown in Listing 3-1 as prog01.py in the directory /**home/pi/DIP/code/chapter03**.

Listing 3-1. prog01.py

```
from PIL import Image
im = Image.open("/home/pi/DIP/Dataset/4.1.07.tiff")
im.show()
```

Run this code with the following command:

```
python3 prog01.py
```

It will show the image in an **xli** window. This is the simplest Pillow program; it loads an image in a Python variable with the function call open() and displays it with the function call show(). In the first line, we are importing the Image module of Pillow. We will learn more about this module in the next chapter. The standard version of show() is not very efficient, because it saves the image to a temporary file and calls the xv utility to display the image. However, it is handy for the purposes of debugging.

The output is as shown in Figure 3-7.

Figure 3-7. *A simple demo*[5]

Due to the limitation of the function show(), we will use Python's built-in GUI module **Tkinter** to display images whenever needed. Listing 3-2 creates an empty window as shown in Figure 3-8.

---

[5] 4.1.07, 4.1.08 Picture of jelly beans taken at USC. Free to use (http://sipi.usc.edu/database/copyright.php)

Listing 3-2. prog02.py

```
import tkinter as tk
root = tk.Tk()
root.title("Test")
root.mainloop()
```

Figure 3-8. *An empty window*

Let's display an image using Tkinter, as shown in Listing 3-3.

Listing 3-3. prog03.py

```
from PIL import Image, ImageTk
import tkinter as tk
im = Image.open("/home/pi/DIP/Dataset/4.1.07.tiff")
root = tk.Tk()
root.title("Test")
photo = ImageTk.PhotoImage(im)
l = tk.Label(root, image=photo)
l.pack()
l.photo = photo
root.mainloop()
```

This program uses Python's built-in module for GUI, called **Tkinter**. We are importing it in the second line. The ImageTk module provides the functionality to convert the **Pillow** image to a Tk-compatible image with the PhotoImage() function. We create a Tk window with the following statement:

```
root = tk.Tk()
```

The code displays the Pillow image as a label. The method title() sets the title of the image. In the code, the Label() and pack() functions are used to create the image label. The line l.photo() = photo is for Python's garbage collector. root.mainloop() is the main loop for the Tk GUI.

The output is shown in Figure 3-9.

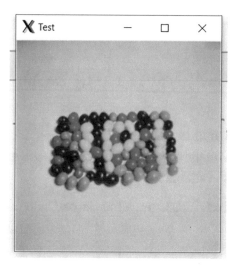

Figure 3-9. *Tkinter and Pillow in action*

Note   If you are encountering an error related to importing while executing this program, run the following command at the terminal: sudo apt-get install python3-pil.imagetk

---

We can also check the properties of an image as follows:

Listing 3-4. prog04.py

```python
from PIL import Image
im = Image.open("/home/pi/DIP/Dataset/4.1.07.tiff")
print(im.mode)
print(im.format)
print(im.size)
print(im.info)
print(im.getbands())
```

It produces the following output:

```
RGB
TIFF
(256, 256)
{'compression': 'raw', 'dpi': (1, 1), 'resolution': (1, 1)}
('R', 'G', 'B')
```

Let's look at the program line by line.

The mode of an image defines the type and depth of its pixels. Here are the standard modes available in Pillow:

- 1 (1-bit pixels, black and white, stored with one pixel per byte)

- L (8-bit pixels, black and white)

- P (8-bit pixels, mapped to any other mode using a color palette)

- RGB (3x8-bit pixels, true color)

- RGBA (4x8-bit pixels, true color with transparency mask)

- CMYK (4x8-bit pixels, color separation)

- YCbCr (3x8-bit pixels, color video format)

The format of the image refers to the file format. Size refers to the resolution of the image in pixels. Info refers to the auxiliary information of the image. The function `getbands()` retrieves the bands of the colors in the image.

Summary

In this chapter, we got started with the basics of Pillow. We learned how to load and display an image. We learned a bit about image properties. We also learned how to capture images using various sensors. In the next chapter, we will explore the modules `Image`, `ImageChops`, and `ImageOps` in **Pillow** in detail.

CHAPTER 4

Basic Operations on Images

In the previous chapter, we started with using Pillow for image processing. We also used **Tkinter** for displaying images. In this chapter, we will learn various arithmetic and logical operations to use on images. We will explore using **Image**, **ImageChops**, and **ImageOps** modules in Pillow to implement these operations. We will also learn how to use the slide bar in Tkinter to dynamically change the program input.

We will be learning about the following topics:

- Image module

- ImageChops module

- ImageOps module

Image Module

Let's start with the **Image** module. We can perform myriad operations on images with this module.

© Ashwin Pajankar 2022
A. Pajankar, *Raspberry Pi Image Processing Programming*,
https://doi.org/10.1007/978-1-4842-8270-0_4

Image Channels

We can use the routine split() to split an image into its constituent channels. We can also merge various images into a single image with the routine merge(). Listing 4-1 shows a demonstration of this.

Listing 4-1. prog01.py

```python
from PIL import Image, ImageTk
import tkinter as tk
im = Image.open("/home/pi/DIP/Dataset/4.1.07.tiff")
root = tk.Tk()
root.title("RED Channel Demo")
r, g, b = im.split()
photo1 = ImageTk.PhotoImage(r)
l1 = tk.Label(root, image=photo1)
l1.pack()
l1.photo = photo1
photo2 = ImageTk.PhotoImage(Image.merge("RGB", (r, g, b)))
l2 = tk.Label(root, image=photo2)
l2.pack()
l2.photo = photo2
root.mainloop()
```

The output is shown in Figure 4-1.

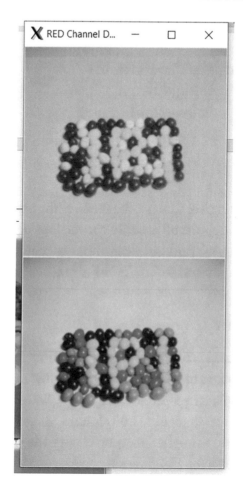

Figure 4-1. *Red channel and the original image*

Colorspace Conversion

You can change the mode of an image using the routine convert(), as shown in Listing 4-2.

Listing 4-2. prog02.py

```
from PIL import Image, ImageTk
import tkinter as tk
im1 = Image.open("/home/pi/DIP/Dataset/4.1.07.tiff")
```

```
res1 = im1.convert("L")
root = tk.Tk()
root.title("Colorspace Conversion Demo")
photo = ImageTk.PhotoImage(res1)
l = tk.Label(root, image=photo)
l.pack()
l.photo = photo
root.mainloop()
```

The code in Listing 4-2 changes the mode of the image to L. The routine convert() supports all possible conversions between the RGB, CMYK, and L modes. We can read more about the colorspaces at https://www.color-management-guide.com/color-spaces.html.

Image Blending

We can blend two images using the blend() method. It takes three arguments—two images to be blended and the value of alpha. The mathematical formula it uses for blending is as follows:

$$\text{output} = \text{image1} * (1.0 - \text{alpha}) + \text{image2} * \text{alpha}$$

Now, we will write a program that can change the value of alpha so that we can experience the blending effect ourselves. We will use the scale widget in Tkinter.

The program in Listing 4-3 demonstrates this process.

Listing 4-3. prog03.py

```
from PIL import Image, ImageTk
import tkinter as tk
def show_value_1(alpha):
    print('Alpha: ', alpha)
    img = Image.blend(im1, im2, float(alpha))
```

```
    photo = ImageTk.PhotoImage(img)
    l['image'] = photo
    l.photo = photo
root = tk.Tk()
root.title('Blending Demo')
im1 = Image.open("/home/pi/DIP/Dataset/4.1.07.tiff")
im2 = Image.open("/home/pi/DIP/Dataset/4.1.08.tiff")
photo = ImageTk.PhotoImage(im1)
l = tk.Label(root, image=photo)
l.pack()
l.photo = photo
w1 = (tk.Scale(root, label="Alpha", from_=0, to=1,
      resolution=0.01, command=show_value_1, orient=tk.
      HORIZONTAL))
w1.pack()
root.mainloop()
```

The code in Listing 4-3 creates a scale widget using the `tk.Scale()` call statement. The statement has more than 79 characters, so in order to conform to the PEP8 standard, I wrote it to fit in two lines, each consisting of fewer than 79 characters. The parameters passed to the `tk.Scale()` call are as follows:

- The Tkinter window variable name

- `label`: The label to be associated with scale

- `from_` and `to`: The range of values

- `resolution`: The resolution of the scale bar

- `command`: The function to execute when the value of the scale changes

- `orient`: The orientation of the scale

When you change the slider with the mouse pointer, it calls the custom show_value_1() function. We are printing the current value of the track-bar position to the console for debugging purposes. The statement img = Image.blend(im1, im2, float(alpha)) creates a blended image. The following lines update the image in the Tkinter window:

```
photo = ImageTk.PhotoImage(img)
l['image'] = photo
l.photo = photo
```

The show_value_1() function updates every time we change the slider position and the new image is computed. This makes the program interesting and interactive, as we can change the value of alpha to see the transition effect.

The output is shown in Figure 4-2.

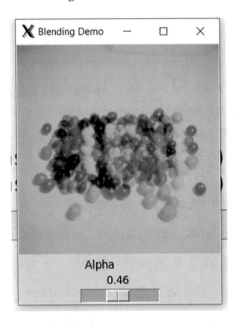

Figure 4-2. *Image blending tool*[1]

---

[1] 4.1.07, 4.1.08 Picture of jelly beans taken at USC. Free to use (http://sipi.usc.edu/database/copyright.php)

We will use the same basic code skeleton to demonstrate the other features of the Pillow library.

Resizing an Image

You can resize an imageusing the routine resize(), as shown in Listing 4-4.

Listing 4-4. prog04.py

```python
from PIL import Image, ImageTk
import tkinter as tk
def show_value_1(size):
    print('Resize: ', size, ' : ', size)
    img = im.resize((int(size), int(size)))
    photo = ImageTk.PhotoImage(img)
    l['image'] = photo
    l.photo = photo
root = tk.Tk()
root.attributes('-fullscreen', True)
im = Image.open("/home/pi/DIP/Dataset/4.1.07.tiff")
photo = ImageTk.PhotoImage(im)
l = tk.Label(root, image=photo)
l.pack()
l.photo = photo
w1 = (tk.Scale(root, label="Resize", from_=128,
      to=512, resolution=1, command=show_value_1, orient=tk.
      HORIZONTAL))
w1.pack()
root.mainloop()
```

This example varies the image size from (128, 128) to (512, 512). The routine `resize()` takes the new size tuple as an argument. The code also invokes the Tkinter window in full-screen mode with the `root.attributes()` function call. To close this window, you have to press Alt+F4 from the keyboard. Run the code and have a look at the output.

Rotating an Image

We can use the routine `rotate()`, which takes the angle of rotation as an argument. The code in Listing 4-5 demonstrates this idea.

Listing 4-5. prog05.py

```python
from PIL import Image, ImageTk
import tkinter as tk
def show_value_1(angle):
    print('Angle: ', angle)
    img = im.rotate(float(angle))
    photo = ImageTk.PhotoImage(img)
    l['image'] = photo
    l.photo = photo
root = tk.Tk()
root.title("Rotation Demo")
im = Image.open("/home/pi/DIP/Dataset/4.1.07.tiff")
photo = ImageTk.PhotoImage(im)
l = tk.Label(root, image=photo)
l.pack()
l.photo = photo
w1 = (tk.Scale(root, label="Angle", from_=0, to=90,
       resolution=1, command=show_value_1, orient=tk.
       HORIZONTAL))
w1.pack()
root.mainloop()
```

Run this code to experience the rotation effect on an image. We can also transpose the images using the transpose() function. It takes one of PIL.Image.FLIP_LEFT_RIGHT, PIL.Image.FLIP_TOP_BOTTOM, PIL.Image.ROTATE_90, PIL.Image.ROTATE_180, PIL.Image.ROTATE_270, or PIL.Image.TRANSPOSE as an argument. The code example in Listing 4-6 shows a rotation at 180 degrees.

Listing 4-6. prog06.py

```
from PIL import Image, ImageTk
import tkinter as tk
root = tk.Tk()
root.title("Transpose Demo")
im = Image.open("/home/pi/DIP/Dataset/4.1.07.tiff")
out = im.transpose(Image.ROTATE_180)
photo = ImageTk.PhotoImage(out)
l = tk.Label(root, image=photo)
l.pack()
l.photo = photo
root.mainloop()
```

Also, the following line (prog07.py in the code bundle) will flip the image vertically:

```
out = im.transpose(Image.FLIP_TOP_BOTTOM)
```

Crop and Paste Operations

We can crop a part of an image using the crop() method. It takes an argument of four tuples that specify the coordinates of the box to be cropped from the image. Pillow also has a paste() method to paste a rectangular image to another image. The paste() method takes the image to be pasted and the coordinates as arguments. The program in Listing 4-7 demonstrates how we can rotate a face using a clever combination of `crop()`, `rotate()`, and `paste()`.

Listing 4-7. prog08.py

```python
from PIL import Image
im = Image.open("/home/pi/DIP/Dataset/4.1.07.tiff")
face_box = (100, 100, 150, 150)
face = im.crop(face_box)
rotated_face = face.transpose(Image.ROTATE_180)
im.paste(rotated_face, face_box)
im.show()
```

The output of the code in Listing 4-7 is shown in Figure 4-3.

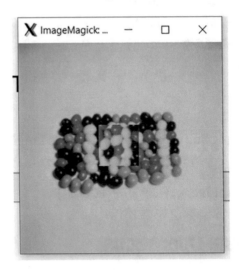

Figure 4-3. *Crop and paste demo*

Copying and Saving Images to a File

You can use the copy() method to copy an entire Pillow image to another Python variable. You can save a Pillow image to a file using the save() method. A demonstration is shown in Listing 4-8.

Listing 4-8. prog09.py

```
from PIL import Image
im = Image.open("/home/pi/DIP/Dataset/4.1.07.tiff")
im1 = im.copy()
im1.save("test.tiff")
```

The code in Listing 4-8 opens an image from a given location, copies it into the im1 variable, and saves it to the current location as test.tiff.

Knowing the Value of a Particular Pixel

You can determine the value of a particular pixel using getpixel(). It is usually a tuple that represents the channel intensities. With RGB images, you get the **Red**, **Green**, and **Blue** intensities. It is used (prog10.py in the code bundle) as follows:

```
print(im.getpixel((100,100)))
```

This way, we can determine the value of the pixel.

Mandelbrot Set

A Mandelbrot set is a fractal, and we can generate it with Pillow.

Note   Read more about Mandelbrot set at https://mathworld. wolfram.com/MandelbrotSet.html

---

We can create a Mandelbrot set with the code in Listing 4-9.

Listing 4-9. prog10_1.py

```python
from PIL import Image
# drawing area (xa < xb and ya < yb)
xa = -2.0
xb = 1.0
ya = -1.5
yb = 1.5
maxIt = 256 # iterations
# image size
imgx = 512
imgy = 512
#create mtx for optimized access
image = Image.new("RGB", (imgx, imgy))
mtx = image.load()
#optimizations
lutx = [j * (xb-xa) / (imgx - 1) + xa for j in range(imgx)]
for y in range(imgy):
    cy = y * (yb - ya) / (imgy - 1)  + ya
    for x in range(imgx):
        c = complex(lutx[x], cy)
        z = 0
        for i in range(maxIt):
            if abs(z) > 2.0: break
            z = z * z + c
        r = i % 4 * 64
        g = i % 8 * 32
        b = i % 16 * 16
        mtx[x, y] =  r,g,b
image.show()
```

```
image.save("mandel.png", "PNG")
```

The output is shown in Figure 4-4.

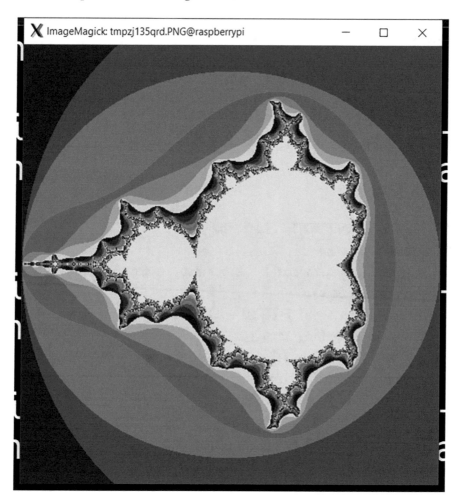

Figure 4-4. *Mandelbrot set*

Julia Set

We can create a Julia set as demonstrated in Listing 4-10.

Listing 4-10. prog10_2.py

```python
from PIL import Image
import random
# drawing area (xa < xb and ya < yb)
xa = -2.0
xb = 1.0
ya = -1.5
yb = 1.5
maxIt = 256 # iterations
# image size
imgx = 512
imgy = 512
image = Image.new("RGB", (imgx, imgy))
# Julia set to draw
c = complex(random.random() * 2.0 - 1.0, random.random() - 0.5)
for y in range(imgy):
    zy = y * (yb - ya) / (imgy - 1)  + ya
    for x in range(imgx):
        zx = x * (xb - xa) / (imgx - 1) + xa
        z = complex(zx, zy)
        for i in range(maxIt):
            if abs(z) > 2.0: break
            z = z * z + c
        r = i % 4 * 64
        g = i % 8 * 32
        b = i % 16 * 16
        image.putpixel((x, y), b * 65536 + g * 256 + r)
image.show()
image.save("julia.png", "PNG")
```

The output is as shown in Figure 4-5.

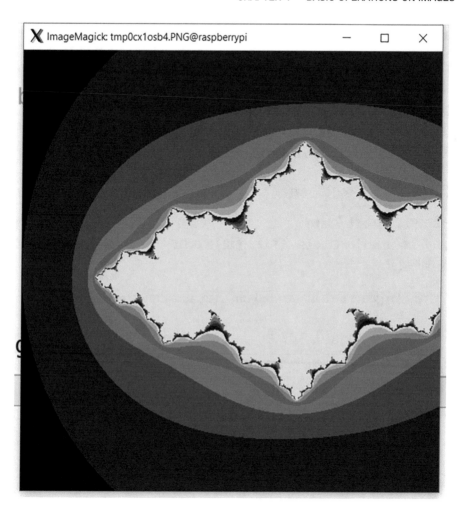

Figure 4-5. *Julia set*

Note   You can read more about Julia sets at the following URLs:

https://mathworld.wolfram.com/JuliaSet.html

https://www.britannica.com/science/Julia-set

Noise and Gradients

Noise is the unwanted part in a signal. Depending on the type of signal, noise can take various forms. We know that an image is a signal. So, an image can have noise. Noise can be inadvertently added into an image at the time of capturing that image or transmission. We can generate (or simulate) noise as shown in Listing 4-11.

Listing 4-11. prog10_3.py

```
from PIL import Image
im = Image.effect_noise((512, 512), 0.01)
im.show()
```

We can generate a linear gradient as demonstrated in Listing 4-12.

Listing 4-12. prog10_4.py

```
from PIL import Image
im = Image.linear_gradient('P')
im.show()
```

We can generate a radial gradient as demonstrated in Listing 4-13.

Listing 4-13. prog10_5.py

```
from PIL import Image
im = Image.radial_gradient('P')
im.show()
```

Please run all these listings to see the code in action.

ImageChops Module

This module has routines for many basic arithmetic and logical operations that we can use on images. Let's look at them quickly one by one.

We can add two images together using the add() method.

The following is sample code for adding images:

```
im3 = ImageChops.add(im1, im2)
```

The add_module() method adds two images without clipping the result:

```
im3 = ImageChops.add_modulo(im1, im2)
```

The darker() method compares two images, pixel by pixel, and returns the darker pixels:

```
im3 = ImageChops.darker(im1, im2)
```

The difference() method returns the difference of the absolute values of two images:

```
im3 = ImageChops.difference(im1, im2)
```

It uses the following mathematical formula for calculating the difference:

$$image3 = abs(image1 - image2)$$

We can invert an image as follows:

```
im2 = ImageChops.invert(im1)
```

Just like with darker(), you can use the lighter() method to return the set of lighter pixels:

```
im3 = ImageChops.lighter(im1, im2)
```

85

logical_and() and logical_or() are the logical operations to use on images. These are explained with the help of black-and-white images. The following are the example usages of these routines:

```
im1 = Image.open("/home/pi/DIP/Dataset/ruler.512.tiff")
im2 = Image.open("/home/pi/DIP/Dataset/ruler.512.tiff")
im2 = im2.transpose(Image.ROTATE_90)
im3 = ImageChops.logical_and(im1.convert("1"), im2.
convert("1"))
im3 = ImageChops.logical_or(im1.convert("1"), im2.convert("1"))
```

These examples convert the grayscale images to black-and-white images first and then perform the logical operations on them. Figure 4-6 shows the output of logical operation AND.

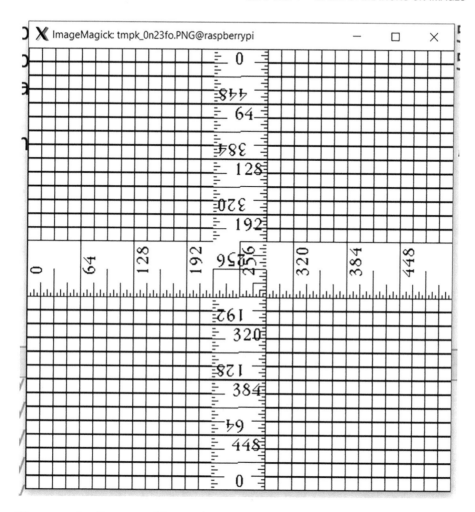

Figure 4-6. *Output of logical operations on black-and-white images*[2]

You can superimpose one image on another using the multiply() method:

```
im3 = ImageChops.multiply(im1, im2)
```

---

[2] gray21.512, ruler.512 Test patterns constructed at USC-SIPI. Free to use. (http://sipi.usc.edu/database/copyright.php)

The `screen()` method superimposes inverted images on top of each other:

```
im3 = ImageChops.screen(im1, im2)
```

You can subtract one image from another using the `subtract()` method as follows:

```
im3 = ImageChops.subtract(im1, im2)
```

You can subtract without clipping the result as follows:

```
im3 = ImageChops.subtract_modulo(im1, im2)
```

You can fill a channel with a given gray level as follows:

```
im2 = ImageChops.constant(im1, 10)
```

You can compute a logical exclusive OR as follows:

```
im3 = ImageChops.logical_xor(im1.convert("1"),
                             im2.convert("1"))
```

You can superimpose two images on top of each other using the Soft Light algorithm as follows:

```
im3 = ImageChops.soft_light(im1, im2)
```

You can superimpose two images on top of each other using the Hard Light algorithm as follows:

```
im3 = ImageChops.hard_light(im1, im2)
```

You can superimpose two images on top of each other using the Overlay algorithm as follows:

```
im3 = ImageChops.overlay(im1, im2)
```

You can also offset the data of an image as follows:

```
im2 = ImageChops.offset(im1,
                        xoffset=128,
                        yoffset=128)
```

ImageOps

This module has many predefined and useful operations.

You can automatically adjust the contrast of an image as follows:

```
im2 = ImageOps.autocontrast(im1)
```

You can colorize a grayscale image as follows:

```
im2 = ImageOps.colorize(im1, (125, 150, 134), (10, 12, 17))
```

You can crop the borders of an image equally from all sides as follows:

```
im2 = ImageOps.crop(im1, 50)
```

The first argument of the ImageOps.crop() method is the image, and the second argument is the width of the cropped border in pixels.

You can also expand the border of an image. Expanded borders will be filled with black pixels equally on all sides, as follows:

```
im2 = ImageOps.expand(im1, 50)
```

You can flip an image vertically as follows:

```
im2 = ImageOps.flip(im1)
```

You can also flip it horizontally as follows:

```
im2 = ImageOps.mirror(im1)
```

You can reduce the number of bits of all the color channels using the posterize() method. This takes the image and the number of bits to keep for every channel as arguments. The following is an example:

```
im2 = ImageOps.posterize(im1, 3)
```

This example keeps only three bits per channel.

The solarize() method inverts all the pixels above a particular grayscale threshold as follows:

```
im2 = ImageOps.solarize(im1, 100)
```

You can pad an image as follows:

```
im2 = ImageOps.pad(im1, (128, 128))
```

You can scale an image as follows:

```
im2 = ImageOps.scale(im1, 1.5)
```

You can automatically equalize the image histogram as follows:

```
im2 = ImageOps.equalize(im1)
```

You can convert a color image to a grayscale image as follows:

```
im2 = ImageOps.grayscale(im1)
```

You can invert an image as follows:

```
im2 = ImageOps.invert(im1)
```

Summary

This chapter explored the modules **Image**, **ImageChops**, and **ImageOps** in detail. In the next chapter, we will explore a few more Pillow modules for advanced operations on images, such as filtering, enhancements, histograms, and quantization.

CHAPTER 5

Advanced Operations on Images

The previous chapter explored the arithmetic and logical operations to use on images with Pillow. There is more to the world of image processing than that, however. Pillow has a lot more functionality to offer. You can enhance and filter images. You can also calculate the histogram of an image as well as its channels. We will cover the following topics in this chapter:

- ImageFilter module

- ImageEnhance module

- Color quantization

- Histograms and equalization

ImageFilter Module

We can use the **ImageFilter** module in Pillow to perform a variety of filtering operations on images. You can use filters to remove noise, to add blur effects, and to sharpen and smooth your images. Listing 5-1 shows a simple image-filtering program that uses the ImageFilter module.

© Ashwin Pajankar 2022
A. Pajankar, *Raspberry Pi Image Processing Programming,*
https://doi.org/10.1007/978-1-4842-8270-0_5

Listing 5-1. prog01.py

```
from PIL import Image, ImageFilter
im1 = Image.open("/home/pi/DIP/Dataset/4.1.08.tiff")
im2 = im1.filter(ImageFilter.BLUR)
#im2 = im1.filter(ImageFilter.CONTOUR)
#im2 = im1.filter(ImageFilter.DETAIL)
#im2 = im1.filter(ImageFilter.EDGE_ENHANCE)
#im2 = im1.filter(ImageFilter.EDGE_ENHANCE_MORE)
#im2 = im1.filter(ImageFilter.EMBOSS)
#im2 = im1.filter(ImageFilter.FIND_EDGES)
#im2 = im1.filter(ImageFilter.SMOOTH)
#im2 = im1.filter(ImageFilter.SMOOTH_MORE)
#im2 = im1.filter(ImageFilter.SHARPEN)
im2.show()
```

Figure 5-1 shows the output of this program.

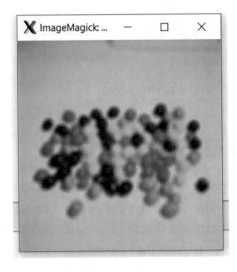

Figure 5-1. *Blur filter demo*

Uncomment the lines in the code file and see the other filters in action.

We can also have a custom filter, as shown in Listing 5-2.

Listing 5-2. prog02.py

```
from PIL import Image, ImageFilter
im1 = Image.open("/home/pi/DIP/Dataset/4.1.08.tiff")
custom_filter = ImageFilter.GaussianBlur(radius=float(5))
im2 = im1.filter(custom_filter)
im2.show()
```

Figure 5-2 shows the output of this program.

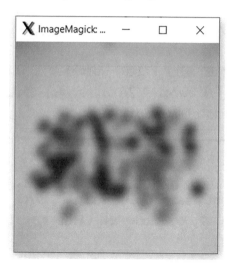

Figure 5-2. *Custom Gaussian filter*

The code in Listing 5-2 uses the GaussianBlur() method for a custom filter. The radius of the blur is 5. We can change the radius. Let's use a slider in **Tkinter** to modify the code and make the blur radius dynamic (see Listing 5-3).

Listing 5-3. prog03.py

```python
from PIL import Image, ImageTk, ImageFilter
import tkinter as tk
def show_value_1(blur_radius):
    print('Gaussian Blur Radius: ', blur_radius)
    custom_filter = ImageFilter.GaussianBlur(radius=float(blur_
    radius))
    img = im1.filter(custom_filter)
    photo = ImageTk.PhotoImage(img)
    l['image'] = photo
    l.photo = photo
root = tk.Tk()
root.title('Gaussian Blur Filter Demo')
im1 = Image.open("/home/pi/DIP/Dataset/4.1.07.tiff")
photo = ImageTk.PhotoImage(im1)
l = tk.Label(root, image=photo)
l.pack()
l.photo = photo
w1 = (tk.Scale(root, label="Blur Radius", from_=0, to=10,
      resolution=0.2, command=show_value_1, orient=tk.
      HORIZONTAL))
w1.pack()
root.mainloop()
```

The output of Listing 5-3 is shown in Figure 5-3.

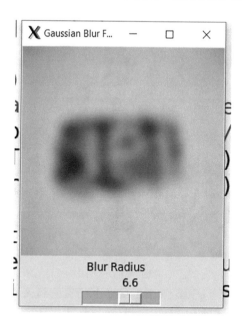

Figure 5-3. *Gaussian blur with TKinter*

We can see the GaussianBlur() has been applied on the image.

The next filter we will explore is the convolution filter. We need to understand the concept of kernels for this example. Kernels are the square matrices used in image processing operations. We mostly use kernels for filtering. Kernels produce a variety of effects, including blurring, smoothing, noise reduction, sharpening, and edge highlighting.

When filters are used to remove high-frequency components from an image, the process is called low-pass filtering, since it allows the low-frequency data to pass through the filter. It creates effects like blurring, smoothing, and noise reduction.

When filters are used to remove low-frequency components from an image, the process is called high-pass filtering, since it allows the high-frequency data to pass through the filter. It creates effects like sharpening and edge highlighting.

Figure 5-4 shows an example kernel.

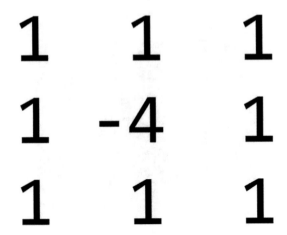

Figure 5-4. *A kernel*

Listing 5-4 shows a code example of the kernel used for image convolution.

Listing 5-4. prog04.py

```
from PIL import Image, ImageTk, ImageFilter
import tkinter as tk
root = tk.Tk()
root.title('Convolution Kernel Demo')
im1 = Image.open("/home/pi/DIP/Dataset/4.1.07.tiff")
custom_filter = ImageFilter.Kernel((3, 3), [1, 1, 1, 1, -4, 1,
1, 1, 1])
img = im1.filter(custom_filter)
photo = ImageTk.PhotoImage(img)
l = tk.Label(root, image=photo)
l.pack()
l.photo = photo
root.mainloop()
```

Run the program in Listing 5-4. It produces the following (Figure 5-5) output.

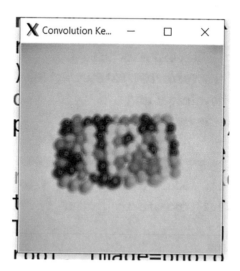

Figure 5-5. *A convolution demo*

As of now, the `ImageFilter.Kernel()` method only supports the kernels of sizes 3x3 and 5x5. Try the following 5x5 kernel:

```
custom_filter = ImageFilter.Kernel((5, 5), [1,1,1,1,1,
1,1,1,1,1, 1,1,-10,1,1, 1,1,1,1,1, 1,1,1,1,1])
```

We can use the Digital Unsharp Mask filter, as shown in Listing 5-5.

Listing 5-5. prog05.py

```
from PIL import Image, ImageTk, ImageFilter
import tkinter as tk
def show_value_1(blur_radius):
    print('Unsharp Blur Radius: ', blur_radius)
    custom_filter = ImageFilter.UnsharpMask(radius=float(blur_
    radius))
    img = im1.filter(custom_filter)
```

```
    photo = ImageTk.PhotoImage(img)
    l['image'] = photo
    l.photo = photo
root = tk.Tk()
root.title('Digital Unsharp Mask Demo')
im1 = Image.open("/home/pi/DIP/Dataset/4.1.07.tiff")
photo = ImageTk.PhotoImage(im1)
l = tk.Label(root, image=photo)
l.pack()
l.photo = photo
w1 = (tk.Scale(root, label="Blur Radius", from_=0, to=10,
    resolution=0.2, command=show_value_1, orient=tk.
    HORIZONTAL))
w1.pack()
root.mainloop()
```

The digital unsharpening mask sharpens the image. Figure 5-6 shows the output with a mask radius of size 10.

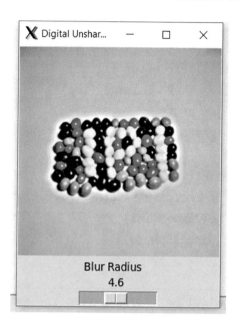

Figure 5-6. *Digital unsharp demo*

The radius of size 10 creates a highly sharpened image.

Let's study the median, min, and max filters in Pillow. All these filters only accept an odd number as the window size. The code in Listing 5-6 demonstrates a median filter.

Listing 5-6. prog06.py

```
from PIL import Image, ImageTk, ImageFilter
import tkinter as tk
def show_value_1(window_size):
    print('Window Size: ', window_size)
    if (int(window_size) % 2 == 0):
        print("Invalid Window Size...\nWindow Size must be an
        odd number...")
    else:
```

```
        custom_filter = ImageFilter.
        MedianFilter(size=int(window_size))
        img = im1.filter(custom_filter)
        photo = ImageTk.PhotoImage (img)
        l['image'] = photo
        l.photo = photo
root = tk.Tk()
root.title('Median Filter Demo')
im1 = Image.open("/home/pi/DIP/Dataset/4.1.07.tiff")
photo = ImageTk.PhotoImage(im1)
l = tk.Label(root, image=photo)
l.pack()
l.photo = photo
w1 = (tk.Scale(root, label="Window Size", from_=1, to=19,
      resolution=1, command=show_value_1, orient=tk.
      HORIZONTAL))
w1.pack()
root.mainloop()
```

The output shown in Figure 5-7 has a window size of 19.

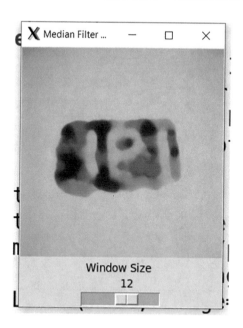

Figure 5-7. *A median filter*

If you change the filter to a min filter using the following code line

```
custom_filter = ImageFilter.MinFilter(size=int(window_size))
```

you'll get the output shown in Figure 5-8.

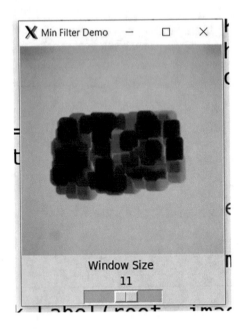

Figure 5-8. *A min filter*

Here's an example of the max filter:

```
custom_filter = ImageFilter.MaxFilter(size=int(window_size))
```

The output for the max filter is shown in Figure 5-9.

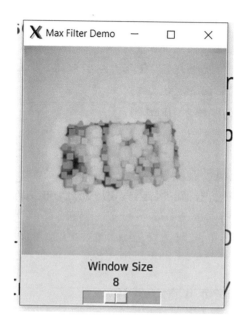

Figure 5-9. *The max filter*

Next, we will see an example of a mode filter. The mode filter works with even and odd window sizes. Listing 5-7 shows a code example of the mode filter.

Listing 5-7. prog09.py

```python
from PIL import Image, ImageTk, ImageFilter
import tkinter as tk
def show_value_1(window_size):
    print('Window Size: ', window_size)
    custom_filter = ImageFilter.
    ModeFilter(size=int(window_size))
    img = im1.filter(custom_filter)
    photo = ImageTk.PhotoImage(img)
    l['image'] = photo
    l.photo = photo
```

```
root = tk.Tk()
root.title('Mode Filter Demo')
im1 = Image.open("/home/pi/DIP/Dataset/4.1.07.tiff")
photo = ImageTk.PhotoImage(im1)
l = tk.Label(root, image=photo)
l.pack()
l.photo = photo
w1 = (tk.Scale(root, label="Window Size", from_=1, to=19,
      resolution=1, command=show_value_1, orient=tk.
      HORIZONTAL))
w1.pack()
root.mainloop()
```

The output is shown in Figure 5-10.

Figure 5-10. *The mode filter*

We can also use the code for BoxBlur() as follows:

```
custom_filter = ImageFilter.BoxBlur(radius=int(window_size))
```

The output is shown in Figure 5-11.

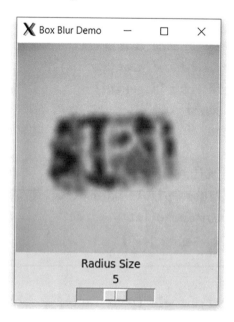

Figure 5-11. *The box blur*

You've seen how to utilize filters; now, let's take a look at how images can be enhanced.

The ImageEnhance Module

We can use the **ImageEnhance** module in Pillow to adjust the contrast, color, sharpness, and brightness of an image, just like we used to do in old analog television sets.

Listing 5-8 shows the code for color adjustment.

Listing 5-8. prog11.py

```python
from PIL import Image, ImageTk, ImageEnhance
import tkinter as tk
def show_value_1(factor):
    print('Color Factor: ', factor)
    enhancer = ImageEnhance.Color(im1)
    img = enhancer.enhance(float(factor))
    photo = ImageTk.PhotoImage(img)
    l['image'] = photo
    l.photo = photo
root = tk.Tk()
root.title('Color Adjustment Demo')
im1 = Image.open("/home/pi/DIP/Dataset/4.1.07.tiff")
photo = ImageTk.PhotoImage(im1)
l = tk.Label(root, image=photo)
l.pack()
l.photo = photo
w1 = (tk.Scale(root, label="Color Factor", from_=0, to=2,
        resolution=0.1, command=show_value_1, orient=tk.
        HORIZONTAL))
w1.pack()
w1.set(1)
root.mainloop()
```

In Listing 5-8, the image processing code is as follows:

```python
enhancer = ImageEnhance.Color(im1)
img = enhancer.enhance(float(factor))
```

We can follow the same style of coding for all the other image enhancement operations. First, we create an enhancer, and then we apply the enhancement factor to that. We also must have observed w1.set(1) in the code. This sets the scale to 1 at the beginning. Changing the argument to set() changes the default position of the scale pointer.

Run the program in Listing 5-8 and take a look at the output.

To change the contrast, use the code in Listing 5-9.

Listing 5-9. prog12.py

```python
from PIL import Image, ImageTk, ImageEnhance
import tkinter as tk
def show_value_1(factor):
    print('Contrast Factor: ', factor)
    enhancer = ImageEnhance.Contrast(im1)
    img = enhancer.enhance(float(factor))
    photo = ImageTk.PhotoImage(img)
    l['image'] = photo
    l.photo = photo
root = tk.Tk()
root.title('Contrast Adjustment Demo')
im1 = Image.open("/home/pi/DIP/Dataset/4.1.07.tiff")
photo = ImageTk.PhotoImage(im1)
l = tk.Label(root, image=photo)
l.pack()
l.photo = photo
w1 = (tk.Scale(root, label="Contrast Factor", from_=0, to=2,
        resolution=0.1, command=show_value_1, orient=tk.
        HORIZONTAL))
w1.pack()
w1.set(1)
root.mainloop()
```

Run the program in Listing 5-9 and take a look at the output.

The following enhancer is used to change the brightness:

```
enhancer = ImageEnhance.Brightness(im1)
```

The following enhancer is used to change the sharpness:

```
enhancer = ImageEnhance.Sharpness(im1)
```

For finer control of the sharpness, use the following code for the scale:

```
w1 = (tk.Scale(root, label="Sharpness Factor", from_=0, to=2,
        resolution=0.1, command=show_value_1, orient=tk.
        HORIZONTAL))
```

These were all image enhancement operations. The next section looks at more advanced image operations.

Color Quantization

Color quantization is the process of reducing the number of distinct colors in an image. The new image should be similar to the original image in appearance. Color quantization is done for a variety of purposes, including when you want to store an image in a digital medium. Real-life images have millions of colors. However, encoding them in the digital format and retaining all the color-related information would result in a huge image size. If you limit the number of colors in the image, you'll need less space to store the color-related information. This is the practical application of quantization. The Image module has a method called quantize() that's used for image quantization.

The code in Listing 5-10 demonstrates image quantization in Pillow.

Listing 5-10. prog15.py

```
from PIL import Image, ImageTk
import tkinter as tk
def show_value_1(num_of_col):
```

```
      print('Number of colors: ', num_of_col)
      img = im1.quantize(int(num_of_col))
      photo = ImageTk.PhotoImage(img)
      l['image'] = photo
      l.photo = photo
root = tk.Tk()
root.title('Color Quantization Demo')
im1 = Image.open("/home/pi/DIP/Dataset/4.1.07.tiff")
photo = ImageTk.PhotoImage(im1)
l = tk.Label(root, image=photo)
l.pack()
l.photo = photo
w1 = (tk.Scale(root, label="Number of colors", from_=4, to=16,
      resolution=1, command=show_value_1, orient=tk.
      HORIZONTAL))
w1.pack()
w1.set(256)
root.mainloop()
```

The following output (Figure 5-12) shows a quantized image.

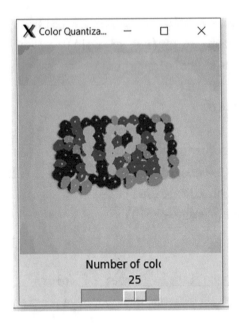

Figure 5-12. *A quantized image*

Histograms and Equalization

You likely studied frequency tables in statistics in school. Well, the histogram is nothing but a frequency table visualized. You can calculate the histogram of any dataset represented in the form of numbers.

The Image module has a method called histogram() that's used to calculate the histogram of an image. An RGB image has three 8-bit channels. This means that it can have a staggering 256 x 256 x 256 number of colors. Drawing a histogram of such a dataset would be very difficult. So, the histogram() method calculates the histogram of individual channels in an image. Each channel has 256 distinct intensities. The histogram is a list of values for each intensity level of a channel.

The histogram for each channel has 256 numbers in the list. Suppose the histogram for the **Red** channel has the values (4, 8, 0, 19, …, 90). This means that four pixels have the red intensity of 0, eight pixels have the red intensity of 1, no pixel has red intensity of 2, 19 pixels have the red intensity of three, and so on, until the last value, which indicates that 90 pixels have the red intensity of 255.

When we combine the histogram of all three channels, we get a list of 768 numbers. In this chapter, we will just compute the histogram. We will not show it visually. When we learn about the advanced image processing library scipy.ndimage, we will learn how to represent histograms for each channel individually.

The code in Listing 5-11 calculates and stores the histograms of an image and its individual channels.

Listing 5-11. prog16.py

```
from PIL import Image
im1 = Image.open("/home/pi/DIP/Dataset/4.1.07.tiff")
print(len(im1.histogram()))
r, g, b = im1.split()
print(len(r.histogram()))
print(len(g.histogram()))
print(len(b.histogram()))
```

Modify this program to directly print the histograms of the image and channels.

A grayscale image (L mode image) will have a histogram of only 256 values because it has only a single channel.

Histogram Equalization

You can adjust the histogram to enhance the image contrast. This technique is known as histogram equalization. The ImageOps.equalize() method equalizes the histogram of the image. Listing 5-12 shows an example of this process.

Listing 5-12. prog17.py

```
from PIL import Image, ImageOps
im1 = Image.open("/home/pi/DIP/Dataset/4.1.07.tiff")
print(im1.histogram())
im2 = ImageOps.equalize(im1)
print(im2.histogram())
im2.show()
```

The program in Listing 5-12 prints the histogram of the original image after the equalization. Add the `im1.show()` statement to the program and then run it to see the difference between the images.

Summary

In this chapter, we explored how to use the Pillow library for advanced image processing. Pillow is good for the beginners who want to get started with an easy-to-program and less mathematical image processing library. However, if you want a more mathematical and scientific approach, then Pillow might not be your best choice. In the following chapters, we will learn about a more powerful library to use for image processing, scipy. ndimage. It is widely used by the scientific community all over the world. We will also learn the basics of the NumPy and matplotlib libraries, which come in handy when processing and displaying images.

CHAPTER 6

Introduction to the Scientific Python Ecosystem

In the previous chapter, we studied advanced image processing with Pillow. Pillow is a nice starting point for image processing operations. However, it has its limitations. When it comes to implementing elaborate image processing operations, such as segmentation, morphological operations, advanced filters, and measurements, Pillow proves to be inadequate. We really need to use better libraries for advanced image processing. The SciPy ecosystem serves as a foundation for all the scientific uses of Python.

SciPy stands for **Scientific Python**. It extensively uses NumPy (Numerical Python) and matplotlib for numeric operations and data visualization, respectively. This chapter covers the basics of NumPy, SciPy, and matplotlib. It also explores introductory-level programming examples using NumPy, scipy.misc, and matplotlib. We will explore the following topics in this chapter:

© Ashwin Pajankar 2022
A. Pajankar, *Raspberry Pi Image Processing Programming*,
https://doi.org/10.1007/978-1-4842-8270-0_6

- The Scientific Python ecosystem

- matplotlib

- Conversion between PIL image objects and NumPy ndarrays

By the end of this chapter, we will be comfortable with the basics of the SciPy ecosystem.

The Scientific Python Ecosystem

The Scientific Python ecosystem is an open source and free collection of libraries for Python to be used in the scientific domain.

It has modules for numerical operations (NumPy), scientific operations (SciPy), and data visualization (matplotlib). It is used extensively by scientific, mathematical, engineering, and analytics communities around the world.

The ecosystem has the following core components:

- NumPy

- SciPy

- matplotlib

- IPython

- Pandas

- Sympy

- Nose

Many other libraries use the core modules in the ecosystem for additional algorithms and data structures. Examples include OpenCV and Scikit.

Figure 6-1 aptly summarizes the role of the Scientific Python ecosystem in the world of scientific computing.

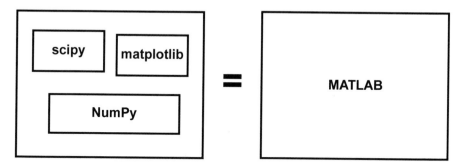

Figure 6-1. *Core components of the Scientific Python ecosystem*

The best way to install the components of the Scientific Python ecosystem on a Raspberry Pi is to use the **pip** utility.

First, upgrade the Pi with the following command:

```
sudo python3 -m pip install --upgrade pip
```

Then, install the components of the Scientific Python ecosystem with the following command:

```
pip3 install numpy scipy matplotlib ipython jupyter pandas
sympy nose
```

This installs all the required libraries.

Simple Examples

Let's study a few simple examples. Let's learn how to read a built in image from **scipy.misc**, store it in a variable, and see the data type of the variable as shown in Listing 6-1,

Listing 6-1. prog01.py

```
from scipy import misc
img = misc.face()
print(type(img))
```

The output is as follows:

```
<class 'numpy.ndarray'>
```

All the variables that store image data belong to this type as far as Scientific Python is concerned. This means that the data type of the image is **ndarray** in **NumPy**. To get started with scientific image processing, and any type of scientific programming in general, we must know what **NumPy** is.

The NumPy homepage at www.numpy.org says this:

NumPy is the fundamental package for scientific computing with Python.

It offers the following features:

- *A powerful multi-dimensional array object*

- *Useful methods for mathematical computations*

- *Wrappers and tools for integration with faster C/C++ and FORTRAN code*

To get started with image processing using SciPy and NumPy, we need to learn the basics of **N-dimensional** (or **multi-dimensional**) array objects in NumPy.

NumPy's N-dimensional array is a homogeneous (contains elements all of the same data type) multi-dimensional array. It has multiple dimensions. Each dimension is known as an axis. The class corresponding to the N-dimensional array in NumPy is numpy.ndarray. This is what we saw in the output of the Listing 6-1. All the major image processing and computer vision libraries, like **Mahotas, OpenCV, scikit-image**, and **scipy.ndimage** (we will study this last one extensively in this book), use numpy.ndarray to represent images. All these libraries have read(), open(), and imread() methods for loading images from disk to a numpy. ndarray object.

NumPy and N-dimensional arrays in NumPy are such vast topics themselves that it would require volumes of books to explain them fully. Hence, we will learn the relevant and important features of these as and when they're needed. For now, you need to understand a few important ndarray properties that will help you comprehend important attributes of the images that ndarray represents.

We can also check the properties of an image. Consider the code in Listing 6-2.

Listing 6-2. prog02.py

```
from scipy import misc
img = misc.face()
print(img.dtype)
print(img.shape)
print(img.ndim)
print(img.size)
```

The output is as follows:

```
uint8
(768, 1024, 3)
3
2359296
```

Let's look at what each of these means. The dtype attribute is for the data type of the elements that represent the image. In this case, it is uint8, which means an unsigned 8-bit integer. This means it can have 256 distinct values. shape means the dimension or size of the images. In this case, it is a color image. Its resolution is 1024 x 768, and it has three color channels corresponding to the colors red, green, and blue. Each channel for each pixel can have one of the 256 possible values. So, a combination can produce 256*256*256 distinct colors for each pixel.

You can visualize a color image as an arrangement of three two-dimensional planes. A grayscale image is a single plane of grayscale values. ndim represents the dimensions. A color image has three dimensions, and a grayscale image has two dimensions. size represents the total number of elements in the array. It can be calculated by multiplying the values of the dimensions. In this case, it is 768*1024*3 = 2359296.

We can see the RGB value corresponding to each individual pixel, as shown in Listing 6-3.

Listing 6-3. prog03.py

```
from scipy import misc
img = misc.face()
print(img[10, 10])
```

The code in Listing 6-3 accesses the value of the pixel located at (10, 10). The output is [172 169 188].

This concludes the basics of NumPy and image processing. You will learn more about NumPy as and when needed throughout the chapters.

Matplotlib

We have thus far used the misc.imshow() method to display an image. While this method is useful for simple applications, it is primitive. We must use a more advanced framework for scientific applications. **Matplotlib** serves this purpose. It is a Matlab-style plotting and data visualization library for Python. We have already installed it. It is an integral part of the Scientific Python ecosystem. Just like NumPy, matplotlib is a vast topic and warrants a dedicated book. The examples in this book use the **pyplot** module in matplotlib for the image processing requirements. Listing 6-4 shows a simple program for image processing.

Listing 6-4. prog04.py

```
import scipy.misc as misc
import matplotlib.pyplot as plt
img = misc.face()
plt.imshow(img)
plt.show()
```

The code in Listing 6-4 imports the **pyplot** module. The imshow() method adds the image to the plot window. The show() method shows the plot window. The output is shown in Figure 6-2.

Figure 6-2. *Matplotlib demo[1]*

We can also turn off the axes (or the ruler) and add a title to the image, as shown in Listing 6-5.

Listing 6-5. prog05.py

```
import scipy.misc as misc
import matplotlib.pyplot as plt
img = misc.ascent()
```

---

[1] Image provided by Judy Weggelaar (This file is in public domain, not copyrighted, no rights reserved, free for any use.) License: https://commons.wikimedia.org/ wiki/File:Raccoon_procyon_lotor.jpg

```
plt.imshow(img, cmap='gray')
plt.axis('off')
plt.title('Ascent')
plt.show()
```

As the image is grayscale, so we have to choose a gray colormap in the imshow() method call so the image colorspace is properly displayed in the plot window. axis('off') is used to turn the axes off. The title() method is used to specify the title of the image. The output is shown in Figure 6-3.

Figure 6-3. *Customizing Matplotlib plot[2]*

---

[2] Image provided by Hillebrand Steve, U.S. Fish and Wildlife Service (This file is in public domain, not copyrighted, no rights reserved, free for any use.) License: https://commons.wikimedia.org/wiki/File:Accent_to_the_top.jpg

We can use imshow() to push multiple images to an image grid in the plot window, as shown in Listing 6-6.

Listing 6-6. prog06.py

```python
import scipy.misc as misc
import matplotlib.pyplot as plt
img1 = misc.face()
img2 = misc.ascent()
titles = ['face', 'ascent']
images = [img1, img2]
plt.subplot(1, 2, 1)
plt.imshow(images[0])
plt.axis('off')
plt.title(titles[0])
plt.subplot(1, 2, 2)
plt.imshow(images[1], cmap='gray')
plt.axis('off')
plt.title(titles[1])
plt.show()
```

We have used the subplot() method before, with imshow(). The first two arguments in the subplot() method specify the dimensions of the grid, and the third argument specifies the position of the image in the grid. The numbering of the images in the grid starts from the top-left edge. The top-left position is the first position, the next position is the second one, and so on. The result is shown in Figure 6-4.

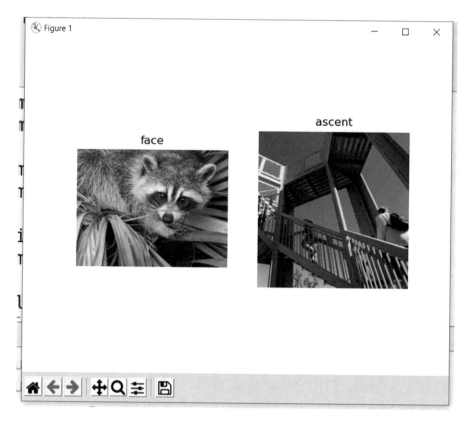

Figure 6-4. *Multiple images in a single plot*

Image Channels

You can separate image channels of a multi-channel image. The code for that process is shown in Listing 6-7.

Listing 6-7. prog07.py

```python
import scipy.misc as misc
import matplotlib.pyplot as plt
img = misc.face()
r = img[:, :, 0]
```

```
g = img[:, :, 1]
b = img[:, :, 2]
titles = ['face', 'Red', 'Green', 'Blue']
images = [img, r, g, b]
plt.subplot(2, 2, 1)
plt.imshow(images[0])
plt.axis('off')
plt.title(titles[0])
plt.subplot(2, 2, 2)
plt.imshow(images[1], cmap='Reds')
plt.axis('off')
plt.title(titles[1])
plt.subplot(2, 2, 3)
plt.imshow(images[2], cmap='Greens')
plt.axis('off')
plt.title(titles[2])
plt.subplot(2, 2, 4)
plt.imshow(images[3], cmap='Blues')
plt.axis('off')
plt.title(titles[3])
plt.show()
```

The result of the code in Listing 6-7 is shown in Figure 6-5.

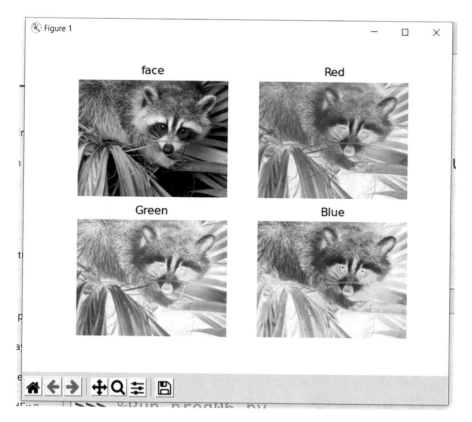

Figure 6-5. *Color channels*

We can use the np.dstack() method, which merges all the channels, to create the original image, as shown in Listing 6-8.

Listing 6-8. prog08.py

```python
import scipy.misc as misc
import matplotlib.pyplot as plt
import numpy as np
img = misc.face()
r = img[:, :, 0]
g = img[:, :, 1]
```

```
b = img[:, :, 2]
output = np.dstack((r, g, b))
plt.imshow(output)
plt.axis('off')
plt.title('Combined')
plt.show()
```

Run the code in Listing 6-8 to see the workings of the np.dstack() for yourself.

Conversion Between PIL Image Objects and NumPy ndarrays

You can use the methods np.asarray() and Image.fromarray() to convert between PIL images and NumPy ndarrays, as shown in Listing 6-9.

Listing 6-9. prog09.py

```
from PIL import Image
import numpy as np
import matplotlib.pyplot as plt
img = Image.open('/home/pi/DIP/Dataset/4.1.07.tiff')
print(type(img))
num_img = np.asarray(img)
print(type(num_img))
img = Image.fromarray(np.uint8(num_img))
print(type(img))
```

The console output is as follows:

```
<class 'PIL.TiffImagePlugin.TiffImageFile'>
<class 'numpy.ndarray'>
<class 'PIL.Image.Image'>
```

Summary

In this chapter, we were introduced to the **Scientific Python** stack and its constituent libraries, such as NumPy and matplotlib. We also explored the **scipy.misc** module for basic image processing and conversion. In the next chapter, we will start exploring the **scipy.ndimage** module for more image processing operations.

CHAPTER 7

Transformations and Measurements

In the previous chapter, we were introduced to the Scientific Python stack. We learned the basics of NumPy and matplotlib. We explored the useful modules **ndarray** and **pyplot** from NumPy and matplotlib, respectively. We also learned about the **scipy.misc** module and how to perform basic image processing with it. In this chapter, we will further explore the SciPy library. We will learn to use the **scipy.ndimage** library for processing images. We will also explore methods to use for image transformation and image measurement.

By the end of this chapter, we will be comfortable with performing operations on images.

Transformations

We studied a few basic transformations that are possible with **scipy.misc** in the previous chapter. Here, we will look at few more.

© Ashwin Pajankar 2022
A. Pajankar, *Raspberry Pi Image Processing Programming*,
https://doi.org/10.1007/978-1-4842-8270-0_7

Note   Just like with **scipy.misc**, **scipy.ndimage** has an `imread()` method that serves the same purpose as `scipy.misc.imread()`. We will be using `face()` and `ascent()` throughout the rest of the book. These are built in library images in SciPy. However, if we want to use other images in the Dataset directory, we can use `imread()` from the **scipy.misc** or **scipy.ndimage** modules in SciPy.

Let's get started with the simple transformation of shifting. The `shift()` method accepts as arguments the image and the values to be applied to the coordinates for shifting. An example is shown in Listing 7-1.

Listing 7-1. prog01.py

```
import scipy.misc as misc
import scipy.ndimage as ndi
import matplotlib.pyplot as plt
img = misc.ascent()
output = ndi.shift(img, 32.0)
plt.imshow(output, cmap='gray')
plt.title('Shift Demo')
plt.axis('off')
plt.show()
```

Figure 7-1 shows the output.

Figure 7-1. *Shift operation demo*

We can also zoom in on the image using the zoom() method. We have to pass the image and the scale of zooming for each of the axes as arguments to the method. Listing 7-2 shows an example of this operation.

Listing 7-2. prog02.py

```
import scipy.misc as misc
import scipy.ndimage as ndi
import matplotlib.pyplot as plt
img = misc.ascent()
plt.imshow(ndi.zoom(img, [5, 3]), cmap='gray')
```

```
plt.title('Zoom Demo')
plt.axis('off')
plt.show()
```

The output of Listing 7-2 is shown in Figure 7-2.

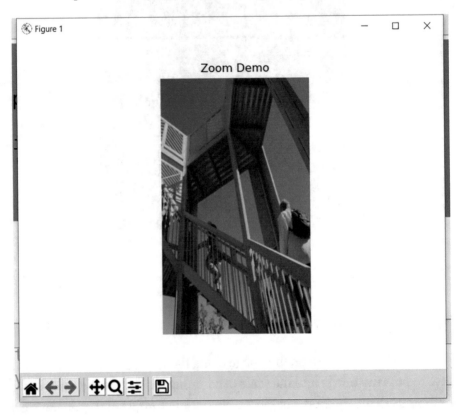

Figure 7-2. *Zoom demo*

We can also use the function rotate() to rotate an image. Listing 7-3 shows several ways to use this function.

Listing 7-3. prog03.py

```python
import scipy.misc as misc
import scipy.ndimage as ndi
import matplotlib.pyplot as plt
img = misc.ascent()
plt.subplot(1, 3, 1)
plt.imshow(img, cmap='gray')
plt.title('Original')
plt.axis('off')
plt.subplot(1, 3, 2)
plt.imshow(ndi.rotate(img, 45, reshape=False), cmap='gray')
plt.title('Not reshaped')
plt.axis('off')
plt.subplot(1, 3, 3)
plt.imshow(ndi.rotate(img, 45, reshape=True), cmap='gray')
plt.title('Reshaped')
plt.axis('off')
plt.show()
```

The output is shown in Figure 7-3.

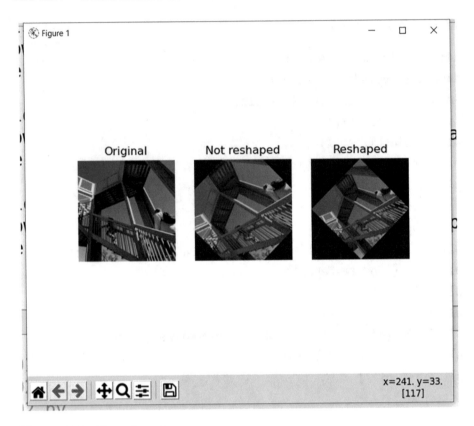

Figure 7-3. *Rotation demo*

We can also apply the shift geometric transformation in another way by defining a custom function. Listing 7-4 demonstrates this.

Listing 7-4. prog04.py

```
import scipy.misc as misc
import scipy.ndimage as ndi
import matplotlib.pyplot as plt
```

```python
def shift_func(output_coords,
                x_shift=128,
                y_shift=128):
    return (output_coords[0] - x_shift,
            output_coords[1] - y_shift)

img = misc.ascent()
plt.subplot(1, 2, 1)
plt.imshow(img, cmap='gray')
plt.title('Original')
plt.subplot(1, 2, 2)
plt.imshow(ndi.geometric_transform(img, shift_func),
cmap='gray')
plt.title('Shift')
plt.show()
```

As seen in Listing 7-4, we pass a customized function for a geometric transformation. The output is as shown in Figure 7-4.

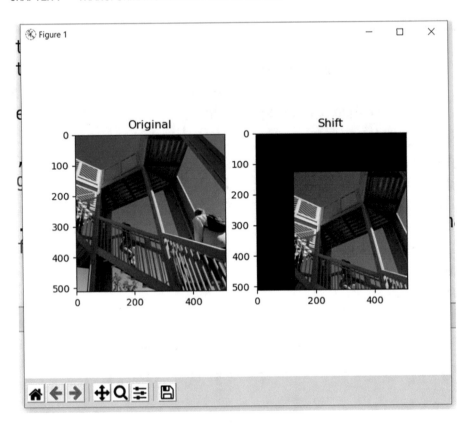

Figure 7-4. *Shifting with geometric transformation*

Let's work with an affine transformation next. An affine transformation is a geometric transformation that preserves the parallelism between lines in the output. Listing 7-5 demonstrates just such a transformation.

Listing 7-5. Affine Transformation

```
import scipy.misc as misc
import scipy.ndimage as ndi
import matplotlib.pyplot as plt
img = misc.ascent()
plt.subplot(1, 3, 1)
```

```python
plt.imshow(img, cmap='gray')
plt.title('Original')
T = [[1, 0, 128], [0, 1, 128], [0, 0, 1]]
plt.subplot(1, 3, 2)
plt.imshow(ndi.affine_transform(img, T), cmap='gray')
plt.title('Translation')
S = [[0.5, 0, 0], [0, 0.5, 0], [0, 0, 1]]
plt.subplot(1, 3, 3)
plt.imshow(ndi.affine_transform(img, S), cmap='gray')
plt.title('Scaling')
plt.show()
```

In an affine transformation, we have to define a transformation matrix and then apply it to the image or geometric shape. In Listing 7-5, we apply scaling and translation (another term for *shifting*) operations to the image. Figure 7-5 shows the output.

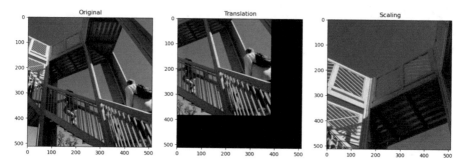

Figure 7-5. *Affine transformations*

We can also apply a spline transformation to an image. We can apply them to an axis individually or to the whole image, as shown in Listing 7-6.

Listing 7-6. Spline Transformation

```python
import scipy.misc as misc
import numpy as np
import scipy.ndimage as ndi
import matplotlib.pyplot as plt
img = misc.ascent()
sp_filter_axis_0 = ndi.spline_filter1d(img, axis=0)
sp_filter_axis_1 = ndi.spline_filter1d(img, axis=1)
sp_filter = ndi.spline_filter(img, order=3)
plt.subplot(2, 2, 1)
plt.imshow(img, cmap='gray')
plt.title('Original')
plt.subplots_adjust(hspace=0.5, wspace=0.5)
plt.subplot(2, 2, 2)
plt.imshow(sp_filter_axis_0, cmap='gray')
plt.title('Spline Filter 1D along X-Axis')
plt.subplot(2, 2, 3)
plt.imshow(sp_filter_axis_1, cmap='gray')
plt.title('Spline Filter 1D along Y-Axis')
plt.subplot(2, 2, 4)
plt.imshow(sp_filter, cmap='gray')
plt.title('Spline Filter')
plt.show()
```

The output is shown in Figure 7-6.

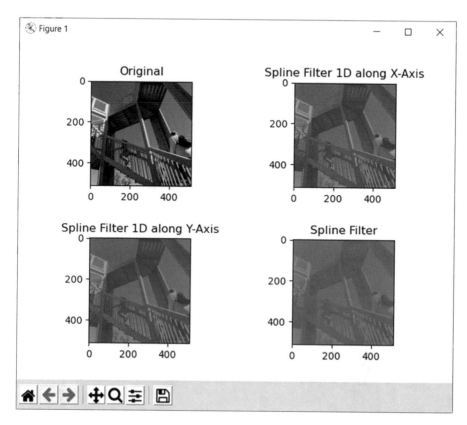

Figure 7-6. *Spline transformations*

Measurements

Let's work with measurements. The first one is the histogram, which is a visual representation of the frequency distribution of pixel intensities. Listing 7-7 shows a demonstration of a histogram.

Listing 7-7. prog07.py

```
import scipy.misc as misc
import scipy.ndimage as ndi
import matplotlib.pyplot as plt
```

139

```
img = misc.face()
hist = ndi.histogram(img, 0, 255, 256)
plt.plot(hist, 'k')
plt.title('Face Histogram')
plt.grid(True)
plt.show()
```

We pass the image, the lowest and highest limits of the bins, and the number of bins as arguments to the function that computes the histogram. The output is shown in Figure 7-7.

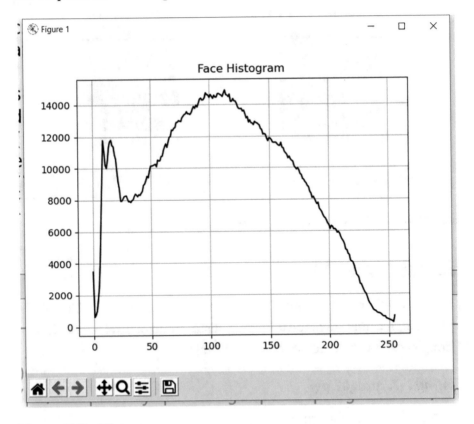

Figure 7-7. *Histogram*

Let's compute a few more measurements, as shown in Listing 7-8.

Listing 7-8. prog08.py

```
import scipy.misc as misc
import scipy.ndimage as ndi
img = misc.ascent()
print(ndi.minimum(img))
print(ndi.minimum_position(img))
print(ndi.maximum(img))
print(ndi.maximum_position(img))
print(ndi.extrema(img))
```

The output is as follows:

```
0
(201, 268)
255
(190, 265)
(0, 255, (201, 268), (190, 265))
```

We can compute a few statistical properties as well, as shown in Listing 7-9.

Listing 7-9. prog09.py

```
import scipy.misc as misc
import scipy.ndimage as ndi
img = misc.ascent()
print(ndi.sum(img))
print(ndi.mean(img))
print(ndi.median(img))
print(ndi.variance(img))
print(ndi.standard_deviation(img))
```

The output is as follows:

```
22932324
87.47987365722656
80.0
2378.9479362999555
48.774459877070456
```

In addition, we can compute the center of mass, as demonstrated in Listing 7-10.

Listing 7-10. prog10.py

```python
import scipy.misc as misc
import scipy.ndimage as ndi
img = misc.ascent()
print(ndi.center_of_mass(img))
```

The output is as follows:

```
(259.9973423539629, 254.6090727219797)
```

Computing the labels in the image can also be achieved, as demonstrated in Listing 7-11.

Listing 7-11. prog11.py

```python
import scipy.misc as misc
import scipy.ndimage as ndi
import numpy as np
img = misc.face()
a = np.zeros((6,6), dtype=int)
a[2:4, 2:4] = 1
a[4, 4] = 1
a[:2, :3] = 2
a[0, 5] = 3
```

```
print(ndi.find_objects(a))
print(ndi.find_objects(a, max_label=2))
print(ndi.find_objects(a, max_label=4))
print(ndi.find_objects(img, max_label=4))
print(ndi.sum_labels(img))
```

The output is as follows:

```
[(slice(2, 5, None), slice(2, 5, None)), (slice(0, 2, None),
slice(0, 3, None)), (slice(0, 1, None), slice(5, 6, None))]
[(slice(2, 5, None), slice(2, 5, None)), (slice(0, 2, None),
slice(0, 3, None))]
[(slice(2, 5, None), slice(2, 5, None)), (slice(0, 2, None),
slice(0, 3, None)), (slice(0, 1, None), slice(5, 6, None)), None]
[(slice(1, 768, None), slice(7, 1024, None), slice(0, 3, None)),
(slice(1, 768, None), slice(8, 1024, None), slice(0, 3, None)),
(slice(1, 768, None), slice(0, 1024, None), slice(0, 3, None)),
(slice(0, 768, None), slice(1, 1023, None), slice(0, 3, None))]
259906521
```

Summary

In this chapter, we explored methods for transformations and measurements. We studied the shift and zoom transformations and calculated the histogram of a grayscale image. We also calculated statistical information about the images.

In the next chapter, we will study image kernels and filters, their types, and their applications in image enhancement in detail.

CHAPTER 8

Filters

In the previous chapter, we learned the functionality found in the **scipy.ndimage** module of the SciPy library. We learned how to apply transformations such as **shift** and **zoom** on an image. We also learned how to obtain statistical information about an image. We saw how to compute and plot the histogram for an image.

In Chapter 5, we studied image filters using the Pillow library. In this chapter, we will study in detail the theory behind those filters. We will also see the types of filters and kernels used in image processing, as well as the applications of those filters.

The following are topics covered in this chapter:

- Kernels

- Low-pass filters

- High-pass filters

- Fourier filters

By the end of this chapter, we will be comfortable with SciPy filters and their applications.

© Ashwin Pajankar 2022
A. Pajankar, *Raspberry Pi Image Processing Programming,*
https://doi.org/10.1007/978-1-4842-8270-0_8

Kernels, Convolution, and Correlation

A kernel (also known as a convolution matrix or a mask) is a two-dimensional matrix used to perform image processing. We can use kernels for operations like convolution and correlation.

Convolution adds each element of the matrix to its local neighbors, weighted by the kernel. Convolution, depending on the kernels used, can produce a variety of results, such as blurring, sharpening, embossing, and edge detection.

While using convolution on matrices, we can use various techniques to handle edges, such as extend, wrap, mirror, crop, kernel crop, constant, and avoid overlap.

Let's demonstrate the functions convolve1d() and convolve() on one-dimensional and two-dimensional matrices, as shown in Listing 8-1.

Listing 8-1. prog00.py

```python
import scipy.ndimage as ndi
import numpy as np
print(ndi.convolve1d([0, 1, 2, 3, 4, 5, 6],
                weights=[1, 2, 3]))
a = np.array([[0, 1, 2, 3],
              [4, 5, 6, 7],
              [8, 9, 10, 11],
              [12, 13, 14, 15]])
k = np.array([[0 ,1, 0],[1, 1, 1],[0, 1, 0]])
print(ndi.convolve(a, k, mode='nearest'))
```

The output is as follows:

```
[ 1   4 10 16 22 28 33]
[[ 5   9 14 18]
 [21 25 30 34]
```

$$\begin{bmatrix} 41 & 45 & 50 & 54 \\ 57 & 61 & 66 & 70 \end{bmatrix}]$$

We can also apply the convolution operation to image matrices. The tutorial at https://setosa.io/ev/image-kernels discusses various kernels and their effects on images. We can use the function convolve2d() from SciPy's **signal** module. Listing 8-2 provides a demonstration.

Listing 8-2. prog01.py

```
import scipy.ndimage as ndi
import scipy.signal
import numpy as np
import scipy.misc as misc
import matplotlib.pyplot as plt
img = misc.ascent()
k = np.ones((11, 11), np.uint8)/121
c1 = ndi.convolve(img, k, mode='nearest')
c2 = scipy.signal.convolve2d(img, k)
print(c1)
print(c2)
plt.subplot(1, 3, 1)
plt.imshow(img, cmap='gray')
plt.title('Original')
plt.subplot(1, 3, 2)
plt.imshow(c1, cmap='gray')
plt.title('Ndimage Convolution')
plt.subplot(1, 3, 3)
plt.imshow(c2, cmap='gray')
plt.title('Signal Convolution')
plt.show()
```

It produces the output shown in Figure 8-1.

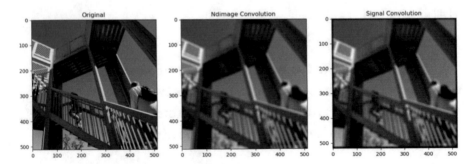

Figure 8-1. *Convolution demo*

Convolution is a mathematical way of generating a single output by combining two inputs. Correlation compares two signals. Listing 8-3 is a demonstration of the correlation operation,

Listing 8-3. prog02.py

```
import scipy.ndimage as ndi
import numpy as np
print(ndi.correlate1d([0, 1, 2, 3, 4, 5, 6],
                  weights=[1, 2, 3]))
a = np.array([[0, 1, 2, 3],
              [4, 5, 6, 7],
              [8, 9, 10, 11],
              [12, 13, 14, 15]])
k = np.array([[0 ,1, 0],[1, 1, 1],[0, 1, 0]])
print(ndi.correlate(a, k, mode='nearest'))
```

The output is as follows:

```
[ 3 8 14 20 26 32 35]
[[ 5 9 14 18]
 [21 25 30 34]
 [41 45 50 54]
 [57 61 66 70]]
```

Let's apply correlation to an image. We will (just like Listing 8-2) use two different methods to compute correlation, as demonstrated in Listing 8-4.

Listing 8-4. prog03.py

```
import scipy.ndimage as ndi
import scipy.signal
import numpy as np
import scipy.misc as misc
import matplotlib.pyplot as plt
img = misc.ascent()
k = np.ones((11, 11), np.uint8)/121
c1 = ndi.correlate(img, k, mode='nearest')
c2 = scipy.signal.correlate2d(img, k)
print(c1)
print(c2)
plt.subplot(1, 3, 1)
plt.imshow(img, cmap='gray')
plt.title('Original')
plt.subplot(1, 3, 2)
plt.imshow(c1, cmap='gray')
plt.title('Ndimage Correlation')
plt.subplot(1, 3, 3)
plt.imshow(c2, cmap='gray')
plt.title('Signal Correlation')
plt.show()
```

The output is as shown in Figure 8-2.

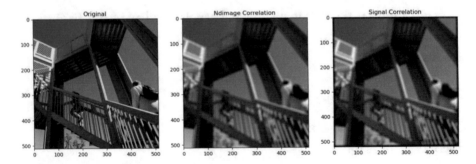

Figure 8-2. *Correlation demo*

Low-Pass Filters

We can use various kernels with the convolution operation to create filters. Filters are used to enhance an image or to perform operations on an image. The filters that remove the high-frequency data from a signal (or an image, for that matter) are known as **low-pass filters**. There are many applications of low-pass filters. Let's look at them now.

Blurring

We can blur images with blurring kernels. SciPy comes with many routines that use blurring kernels with convolution. Listing 8-5 demonstrates a Gaussian filter's being applied to an image.

Listing 8-5. prog04.py

```
import scipy.misc as misc
import scipy.ndimage as ndi
import matplotlib.pyplot as plt
img = misc.face()
output1 = ndi.gaussian_filter(img, sigma=3)
output2 = ndi.gaussian_filter(img, sigma=5)
```

```
output3 = ndi.gaussian_filter(img, sigma=7)
output = [output1, output2, output3]
titles = ['Sigma = 3', 'Sigma = 5',
          'Sigma = 7']
for i in range(3):
    plt.subplot(1, 3, i+1)
    plt.imshow(output[i], cmap='gray')
    plt.title(titles[i])
    plt.axis('off')
plt.show()
```

The routine for the Gaussian filter accepts the image and the value for sigma. The output is as shown in Figure 8-3.

Figure 8-3. *Gaussian filter demo*

A uniform filter replaces the value of a pixel by the mean value of an area centered at the pixel. Let's see a uniform filter, as demonstrated in Listing 8-6.

Listing 8-6. prog05.py

```
import scipy.misc as misc
import scipy.ndimage as ndi
import matplotlib.pyplot as plt
img = misc.face()
output1 = ndi.uniform_filter(img, size=19)
output2 = ndi.uniform_filter(img, size=25)
output3 = ndi.uniform_filter(img, size=31)
```

```
output = [output1, output2, output3]
titles = ['Size = 19', 'Size = 25', 'Size = 31']
for i in range(3):
        plt.subplot(1, 3, i+1)
        plt.imshow(output[i], cmap='gray')
        plt.title(titles[i])
        plt.axis('off')
plt.show()
```

The routine for the uniform filter accepts the image and the size of the filter. The output is as shown in Figure 8-4.

Figure 8-4. *Uniform filter demo*

We can also have a one-dimensional Gaussian filter, as demonstrated in Listing 8-7.

Listing 8-7. prog06.py

```
import scipy.misc as misc
import scipy.ndimage as ndi
import matplotlib.pyplot as plt
img = misc.ascent()
y3 = ndi.gaussian_filter1d(img, sigma=3)
y6 = ndi.gaussian_filter1d(img, sigma=6)
plt.subplot(1, 3, 1)
plt.imshow(img, cmap='gray')
plt.title('Original Image')
```

```
plt.subplot(1, 3, 2)
plt.imshow(y3, cmap='gray')
plt.title('Sigma = 3')
plt.subplot(1, 3, 3)
plt.imshow(y6, cmap='gray')
plt.title('Sigma = 3')
plt.show()
```

The routine for the one-dimensional Gaussian filter accepts the image and the value of Sigma. The output is as shown in Figure 8-5.

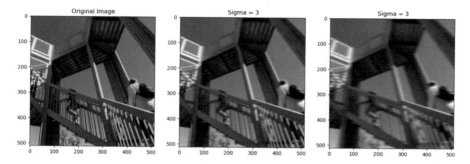

Figure 8-5. *One-dimensional Gaussian filter demo*

We can even apply a one-dimensional uniform filter to an image, as demonstrated in Listing 8-8.

Listing 8-8. prog07.py

```
import scipy.misc as misc
import scipy.ndimage as ndi
import matplotlib.pyplot as plt
img = misc.ascent()
y9 = ndi.uniform_filter1d(img, size=9)
y12 = ndi.uniform_filter1d(img, size=12)
plt.subplot(1, 3, 1)
plt.imshow(img, cmap='gray')
```

```
plt.title('Original Image')
plt.subplot(1, 3, 2)
plt.imshow(y9, cmap='gray')
plt.title('Size = 9')
plt.subplot(1, 3, 3)
plt.imshow(y12, cmap='gray')
plt.title('Size = 12')
plt.show()
```

The routine for the one-dimensional uniform filter accepts the image and the size of the filter. The output is as shown in Figure 8-6.

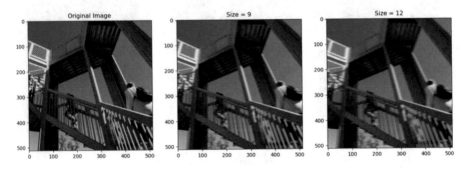

Figure 8-6. *One-dimensional uniform filter demo*

A percentile filter computes the n-percentile value of neighborhood pixel and applies it to the center pixel. We can apply a percentile filter to an image, as demonstrated in Listing 8-9.

Listing 8-9. prog08.py

```
import scipy.misc as misc
import scipy.ndimage as ndi
import matplotlib.pyplot as plt
img = misc.ascent()
out1 = ndi.percentile_filter(img, size=10,
                             percentile=10)
```

```
out2 = ndi.percentile_filter(img, size=45,
                             percentile=65)
plt.subplot(1, 3, 1)
plt.imshow(img, cmap='gray')
plt.title('Original Image')
plt.subplot(1, 3, 2)
plt.imshow(out1, cmap='gray')
plt.title('percentile=10')
plt.subplot(1, 3, 3)
plt.imshow(out2, cmap='gray')
plt.title('percentile=65')
plt.show()
```

The output is as shown in Figure 8-7.

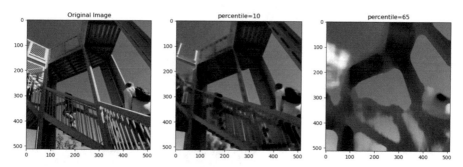

Figure 8-7. *Percentile filter demo*

Rank filters are non-linear filters. A rank filter uses local gray-level ordering to compute the output. We can even apply a rank filter to an image, as demonstrated in Listing 8-10.

Listing 8-10. prog09.py

```
import scipy.misc as misc
import scipy.ndimage as ndi
import matplotlib.pyplot as plt
```

```
img = misc.ascent()
out1 = ndi.rank_filter(img, size=10, rank=10)
out2 = ndi.rank_filter(img, size=45, rank=65)
plt.subplot(1, 3, 1)
plt.imshow(img, cmap='gray')
plt.title('Original Image')
plt.subplot(1, 3, 2)
plt.imshow(out1, cmap='gray')
plt.title('rank=10')
plt.subplot(1, 3, 3)
plt.imshow(out2, cmap='gray')
plt.title('rank=65')
plt.show()
```

The output is as shown in Figure 8-8.

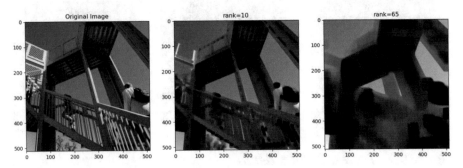

Figure 8-8. *Rank filter demo*

Noise Reduction

We can use low-pass filters for noise reduction. In the next two examples
(Listing 8-11 and Listing 8-12), we will create artificial noise with NumPy
routines and add it to the test image. Listing 8-11 shows how to reduce
noise with the Gaussian filter.

Listing 8-11. prog10.py

```python
import scipy.misc as misc
import scipy.ndimage as ndi
import numpy as np
import matplotlib.pyplot as plt
img = misc.ascent()
noisy = img + 0.8 * img.std() * np.random.random(img.shape)
output1 = ndi.gaussian_filter(noisy, sigma=1)
output2 = ndi.gaussian_filter(noisy, sigma=3)
output3 = ndi.gaussian_filter(noisy, sigma=5)
output = [noisy, output1, output2, output3]
titles = ['Noisy', 'Sigma = 1', 'Sigma = 3', 'Sigma = 5']
for i in range(4):
    plt.subplot(2, 2, i+1)
    plt.imshow(output[i], cmap='gray')
    print(output[i])
    plt.title(titles[i])
    plt.axis('off')
plt.show()
```

The output is as shown in Figure 8-9.

Noisy

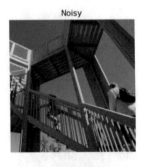

Sigma = 1

Sigma = 3

Sigma = 5

Figure 8-9. *Gaussian filter for noise reduction*

Similarly, we can apply a median filter on a noisy image, as demonstrated in Listing 8-12.

Listing 8-12. prog11.py

```
import scipy.misc as misc
import scipy.ndimage as ndi
import numpy as np
import matplotlib.pyplot as plt
img = misc.ascent()
noisy = img + 0.8 * img.std() * np.random.random(img.shape)
output1 = ndi.median_filter(noisy, 3)
output2 = ndi.median_filter(noisy, 7)
output3 = ndi.median_filter(noisy, 9)
output = [noisy, output1, output2, output3]
```

```
titles = ['Noisy', 'Size = 3', 'Size = 7', 'Size = 9']
for i in range(4):
      plt.subplot(2, 2, i+1)
      plt.imshow(output[i], cmap='gray')
      plt.title(titles[i])
      plt.axis('off')
plt.show()
```

The output is as shown in Figure 8-10.

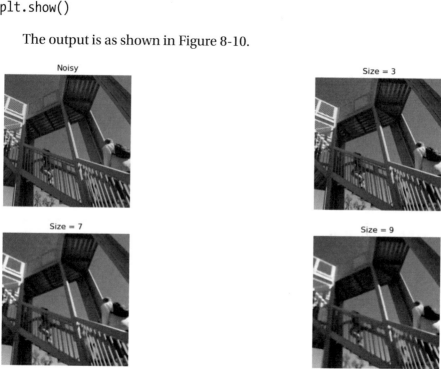

Figure 8-10. *Median filter for noise reduction*

Minimum and maximum filters can also be applied to a noisy image, as demonstrated in Listing 8-13.

Listing 8-13. prog12.py

```
import scipy.misc as misc
import scipy.ndimage as ndi
```

```
import numpy as np
import matplotlib.pyplot as plt
img = misc.ascent()
output1 = ndi.minimum_filter1d(img, 9)
output2 = ndi.minimum_filter(img, 9)
output3 = ndi.maximum_filter1d(img, 9)
output4 = ndi.maximum_filter(img, 9)
output = [output1, output2, output3, output4]
titles = ['Minimum 1D', 'Minimum',
          'Maximum 1D', 'Maximum']
for i in range(4):
    plt.subplot(2, 2, i+1)
    plt.imshow(output[i], cmap='gray')
    plt.title(titles[i])
    plt.axis('off')
plt.show()
```

The output is as shown in Figure 8-11.

Minimum 1D

Minimum

Maximum 1D

Maximum

Figure 8-11. *Minimum and maximum filters*

High-Pass Filters

High-pass filters work in the exact opposite way to the low-pass filters. They allow high-frequency signals to pass through them. They also use kernels and convolutions. However, the kernel used for high-pass filters is different than that used in low-pass filters. When we apply high-pass filters to an image, it sharpens the image (which is opposite of blurring) and highlights edges.

Listing 8-14 demonstrates the Prewitt filter to detect edges in the test image.

Listing 8-14. prog13.py

```python
import scipy.misc as misc
import scipy.ndimage as ndi
import matplotlib.pyplot as plt
img = misc.ascent()
filtered = ndi.prewitt(img)
output = [img, filtered]
titles = ['Original', 'Filtered']
for i in range(2):
    plt.subplot(1, 2, i+1)
    plt.imshow(output[i], cmap='gray')
    plt.title(titles[i])
    plt.axis('off')
plt.show()
```

The output is as shown in Figure 8-12.

Figure 8-12. *Prewitt filter*

We can see the highlighted edges in Figure 8-12.

We can also apply Sobel filters to axes separately and to the entire image. It will detect horizontal edges, vertical edges, and edges on both the axes. Listing 8-15 generates a test image and applies Sobel filters.

Listing 8-15. prog14.py

```python
import numpy as np
import scipy.ndimage as ndi
import matplotlib.pyplot as plt
img = np.zeros((516, 516))
img[128:-128, 128:-128] = 1
img = ndi.gaussian_filter(img, 8)
rotated = ndi.rotate(img, -20)
noisy = rotated + 0.09 * np.random.random(rotated.shape)
sx = ndi.sobel(noisy, axis=0)
sy = ndi.sobel(noisy, axis=1)
sob = np.hypot(sx, sy)
titles = ['Original', 'Rotated', 'Noisy',
          'Sobel (X-axis)', 'Sobel (Y-axis)', 'Sobel']
output = [img, rotated, noisy, sx, sy, sob]
for i in range(6):
    plt.subplot(2, 3, i+1)
    plt.imshow(output[i])
    plt.title(titles[i])
    plt.axis('off')
plt.show()
```

The output is as shown in Figure 8-13.

Figure 8-13. *Sobel filters*

Finally, Listing 8-16 demonstrates the Laplacian, Gaussian gradient magnitude, and Gaussian Laplace filters.

Listing 8-16. progl5.py

```python
import scipy.misc as misc
import scipy.ndimage as ndi
import matplotlib.pyplot as plt
img = misc.ascent()
output1 = ndi.laplace(img)
output2 = ndi.gaussian_gradient_magnitude(img,
                                          sigma=3)
output3 = ndi.gaussian_laplace(img, sigma=3)
titles = ['Original', 'Laplacian',
          'GGM', 'Gaussian Laplace']
output = [img, output1,
          output2, output3]
for i in range(4):
```

```
        plt.subplot(2, 2, i+1)
        plt.imshow(output[i],
                    cmap='gray')
        plt.title(titles[i])
        plt.axis('off')
plt.show()
```

The output is as shown in Figure 8-14.

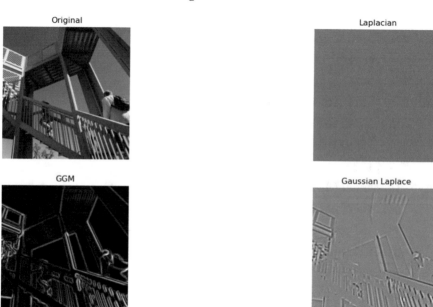

Figure 8-14. *Laplacian, Gaussian gradient magnitude, and Gaussian Laplace filters*

Fourier Filters

Fourier filters work in the frequency domain. They compute the fourier transform of the image, manipulate frequencies, and then finally compute the inverse fourier transform to produce the final output. We can apply various Fourier filtersFiltersfourier filters to images, as demonstrated in Listing 8-17.

Listing 8-17. prog16.py

```python
import scipy.ndimage as ndi
import scipy.misc as misc
import matplotlib.pyplot as plt
import numpy as np
img = misc.ascent()
noisy = img + 0.09 * img.std() * np.random.random(img.shape)
fe = ndi.fourier_ellipsoid(img, 1)
fg = ndi.fourier_gaussian(img, 1)
fs = ndi.fourier_shift(img, 1)
fu = ndi.fourier_uniform(img, 1)
titles = ['Original', 'Noisy',
          'Fourier Ellipsoid', 'Fourier Gaussian',
          'Fourier Shift', 'Fourier Uniform']
output = [img, noisy, fe, fg, fs, fu]
for i in range(6):
    plt.subplot(2, 3, i+1)
    plt.imshow(np.float64(output[i]), cmap='gray')
    plt.title(titles[i])
    plt.axis('off')
plt.show()
```

The output is as shown in Figure 8-15.

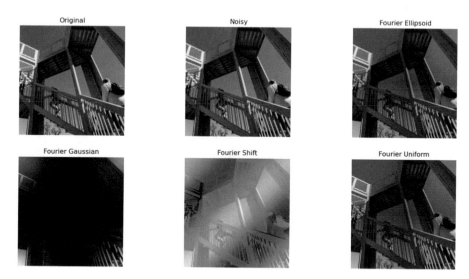

Figure 8-15. *Various Fourier filters*

Summary

In this chapter, we were introduced to myriad filters. We saw their types and looked at their applications. The image-filtering topic is too vast to be completely explored in a single chapter.

In the next chapter, we will cover morphological operators, image thresholding, and basic segmentation.

167

CHAPTER 9

Morphology, Thresholding, and Segmentation

In the previous chapter, we studied the theory behind image filters, along with the types and practical applications of filters used to enhance images.

In this chapter, we are going to study and demonstrate important concepts in image processing, such as morphology, morphological operations on images, thresholding, and segmentation. We will study and demonstrate the following:

- Distance transforms

- Morphology and morphological operations

- Thresholding and segmentation

Distance Transforms

A distance transform is an operation performed on binary images. Binary images have background elements (zero value – black color) and foreground elements (white color). A distance transform replaces

© Ashwin Pajankar 2022
A. Pajankar, *Raspberry Pi Image Processing Programming*,
https://doi.org/10.1007/978-1-4842-8270-0_9

each foreground element with the value of the shortest distance to the
background. **scipy.ndimage** has three methods for computing the
distance transform of a binary image. The code in Listing 9-1 illustrates
how a distance transform can be used practically to generate test images.

Listing 9-1. prog01.py

```python
import scipy.ndimage as ndi
import matplotlib.pyplot as pltimport numpy as np
img = np.zeros((32, 32))
img[8:-8, 8:-8] = 1
print(img)
dist1 = ndi.distance_transform_bf(img)
dist2 = ndi.distance_transform_cdt(img)
dist3 = ndi.distance_transform_edt(img)
output = [img, dist1, dist2, dist3]
titles = ['Original', 'Brute Force', 'Chamfer', 'Euclidean']
for i in range(4):
    print(output[i])
    plt.subplot(2, 2, i+1)
    plt.imshow(output[i],
            interpolation='nearest')
    plt.title(titles[i])
    plt.axis('off')
plt.show()
```

The code shown in Listing 9-1 computes the distance transform by
using a brute-force algorithm, a Chamfer-type algorithm, and Euclidean
methods, respectively. We are going to use distance transforms to generate
the test images in this chapter. The output is shown in Figure 9-1.

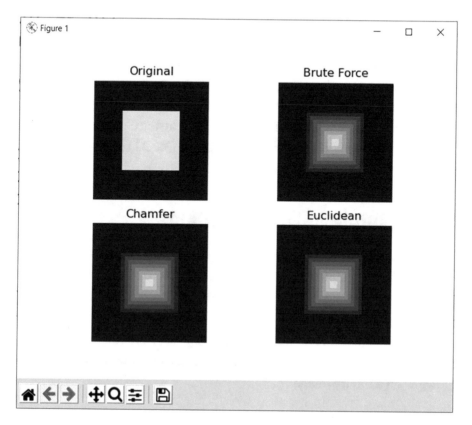

Figure 9-1. *Distance transform*

Morphology and Morphological Operations

Morphology is the study of shapes and forms. The morphological study of images deals with shapes rather than with the values of the pixels. Morphological operations are usually performed on binary images. Let's take a look at a few concepts related to morphological operations.

Structuring Element

A structuring element is a matrix that's used to interact with a given binary image. It comes in various shapes, like a ball, a ring, or a line. It can come in many shapes, like a 3 x 3 or a 7 x 7 matrix. Larger structuring elements take more time for computation. A simple structuring element can be defined as a unity matrix of odd sizes. np.ones((3, 3)) is an example of this.

Binary Morphological Operations

Let's briefly look at various morphological operations. Dilation causes the expansion of the shapes in an input image. Erosion causes shrinkage in the shape in an input image. Opening is the dilation of erosion. Closing is the erosion of dilation.

It is difficult to understand these concepts just by reading about them. Take a look at the example in Listing 9-2, which demonstrates these concepts on a binary image.

Listing 9-2. prog02.py

```
import matplotlib.pyplot as plt
import scipy.ndimage as ndi
import numpy as np
img = np.zeros((16, 16))
img[4:-4, 4:-4] = 1
print(img)
erosion = ndi.binary_erosion(img).astype(img.dtype)
dilation = ndi.binary_dilation(img).astype(img.dtype)
opening = ndi.binary_opening(img).astype(img.dtype)
closing = ndi.binary_closing(img).astype(img.dtype)
output = [img, erosion, dilation,
```

```
            opening, closing]
titles = ['Original', 'Erosion',
          'Dilation', 'Opening',
          'Closing']
for i in range(5):
    print(output[i])
    plt.subplot(1, 5, i+1)
    plt.imshow(output[i],
            interpolation='nearest')
    plt.title(titles[i])
    plt.axis('off')
plt.show()
```

The code example in Listing 9-2 generates a binary image and applies all the binary morphological operations to it. The output is shown in Figure 9-2.

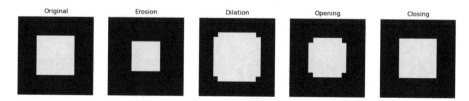

Figure 9-2. *Morphological operations on a binary image*

Another important operation is binary_fill_holes(). It is used to fill the gaps in the binary image, as shown in Listing 9-3.

Listing 9-3. prog03.py

```
import matplotlib.pyplot as plt
import scipy.ndimage as ndi
import numpy as np
img = np.ones((32, 32))
x, y = (32*np.random.random((2, 20))).astype(np.int)
```

```
img[x, y] = 0
noise_removed = ndi.binary_fill_holes(img).astype(int)
output = [img, noise_removed]
titles = ['Original', 'Noise Removed']
for i in range(2):
    print(output[i])
    plt.subplot(1, 2, i+1)
    plt.imshow(output[i],
            interpolation='nearest')
    plt.title(titles[i])
    plt.axis('off')
plt.show()
```

The code in Listing 9-3 generates a 32 x 32 square matrix image with all the values as 1 (white value). Then it randomly sets a few pixels to 0 (the dark value). The dark pixels can be considered holes in the image, which can then be removed using binary_fill_holes().

The output is shown in Figure 9-3.

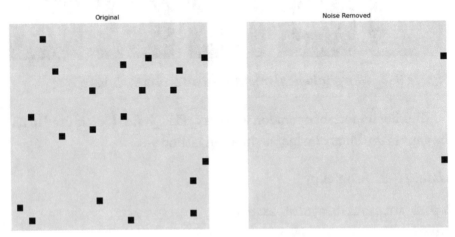

Figure 9-3. *Noise removal with a morphological operation*

Earlier, we discussed binary structure. Let's apply the morphological operations on an image using a binary structure. Listing 9-4 demonstrates this.

Listing 9-4. prog04.py

```python
import matplotlib.pyplot as plt
import scipy.ndimage as ndi
import numpy as np
img = np.zeros((16, 16))
img[4:-4, 4:-4] = 1
struct = ndi.generate_binary_structure(2, 1)
print(struct)
erosion = ndi.binary_erosion(img, struct).astype(img.dtype)
dilation = ndi.binary_dilation(img, struct).astype(img.dtype)
opening = ndi.binary_opening(img, struct).astype(img.dtype)
closing = ndi.binary_closing(img, struct).astype(img.dtype)
output = [img, erosion, dilation,
          opening, closing]
titles = ['Original', 'Erosion',
          'Dilation', 'Opening',
          'Closing']
for i in range(5):
    print(output[i])
    plt.subplot(1, 5, i+1)
    plt.imshow(output[i],
            interpolation='nearest')
    plt.title(titles[i])
    plt.axis('off')
plt.show()
```

The output is as shown in Figure 9-4.

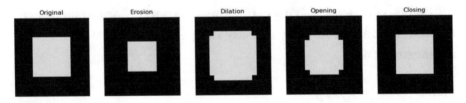

Figure 9-4. *Morphological operations with structuring element*

We can apply a hit or miss operation with and without a structuring element, as demonstrated in Listing 9-5.

Listing 9-5. prog05.py

```python
import matplotlib.pyplot as plt
import scipy.ndimage as ndi
import numpy as np
img = np.zeros((7, 7),
               dtype=int)
img[1, 1] = img[2:4, 2:4] = img[4:6, 4:6] = 1
struct = np.array([[1, 0, 0],
                   [0, 1, 1],
                   [0, 1, 1]])
print(struct)
hit_or_miss1 = ndi.binary_hit_or_miss(img).astype(int)
hit_or_miss2 = ndi.binary_hit_or_miss(img, structure1=struct).
               astype(int)
output = [img, hit_or_miss1, hit_or_miss2]
titles = ['Original',
          'Hit or Miss 1',
          'Hit or Miss 2']
for i in range(3):
    print(output[i])
    plt.subplot(1, 3, i+1)
```

```
        plt.imshow(output[i],
                    interpolation='nearest')
        plt.title(titles[i])
        plt.axis('off')
plt.show()
```

The output is as shown in Figure 9-5.

Figure 9-5. *Binary hit or miss operation*

We can even demonstrate binary propagation, as shown in Listing 9-6.

Listing 9-6. prog06.py

```
import matplotlib.pyplot as plt
import scipy.ndimage as ndi
import numpy as np
n = 10
img = np.zeros((n, n),
                dtype=np.uint8)
img[2, 2] = 1
mask = np.zeros((n, n),
                dtype=np.uint8)
mask[1:4, 1:4] = mask[4, 4] = mask[6:8, 6:8] = 1
struct = np.ones((3, 3))
o1=ndi.binary_propagation(img, mask=mask).astype(int)
```

```
o2=ndi.binary_propagation(img, mask=mask,
                          structure=struct).astype(int)
output = [img, o1, o2]
titles = ['Original',
          'Output 1',
          'Output 2']
for i in range(3):
    print(output[i])
    plt.subplot(1, 3, i+1)
    plt.imshow(output[i],
            interpolation='nearest')
    plt.title(titles[i])
    plt.axis('off')
plt.show()
```

The output is as shown in Figure 9-6.

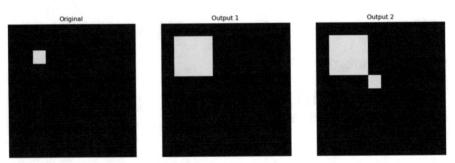

Figure 9-6. *Binary propagation*

Grayscale Morphological Operations

There is a set of grayscale morphological operations. The code shown in Listing 9-7 generates eight random values and assigns them to pixels on a completely dark (zero value) background. Then it uses the grey_ dilation() method to dilate them.

Listing 9-7. prog07.py

```python
import matplotlib.pyplot as plt
import scipy.ndimage as ndi
import numpy as np
img = np.zeros((64, 64))
x, y = (63*np.random.random((2, 8))).astype(np.int)
img[x, y] = np.arange(8)
dilation = ndi.grey_dilation(img, size=(5, 5),
                             structure=np.ones((5, 5)))
output = [img, dilation]
titles = ['Original', 'Dilation']
for i in range(2):
    print(output[i])
    plt.subplot(1, 2, i+1)
    plt.imshow(output[i],
               interpolation='nearest')
    plt.title(titles[i])
    plt.axis('off')
plt.show()
```

The output is shown in Figure 9-7.

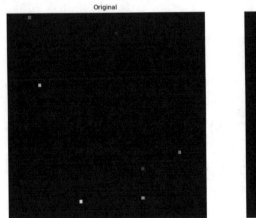
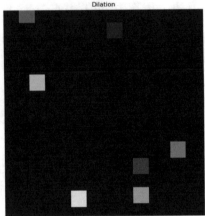

Figure 9-7. *Gray dilation*

We are using a structuring element that's 5 x 5 here.

Note   This example uses the `random()` method. When you will execute the code in the bundle, the output won't be exactly the same.

The code shown in Listing 9-8 applies gray dilation and gray erosion operations to a distance transform.

Listing 9-8. prog08.py

```
import matplotlib.pyplot as plt
import scipy.ndimage as ndi
import numpy as np

img = np.zeros((16, 16))
img[4:-4, 4:-4] = 1

img = ndi.distance_transform_bf(img)
```

```python
dilation = ndi.grey_dilation(img, size=(3, 3),
                             structure=np.ones((3, 3)))

erosion = ndi.grey_erosion(img, size=(3, 3),
                           structure=np.ones((3, 3)))

output = [img, dilation, erosion]
titles = ['Original', 'Dilation', 'Erosion']

for i in range(3):
    print(output[i])
    plt.subplot(1, 3, i+1)
    plt.imshow(output[i], interpolation='nearest',
    cmap='rainbow')
    plt.title(titles[i])
    plt.axis('off')
plt.show()
```

The code in Listing 9-8 uses a structuring element that's 3 x 3 for both operations. The output is shown in Figure 9-8.

Figure 9-8. *Dilation and erosion on a distance transform*

Thresholding and Segmentation

This is the final part of the book and it deals with one of the most important applications of image processing: segmentation. In thresholding operations, you convert grayscale images to binary (black and white) images based on the threshold value. The pixels with intensity values greater than the threshold are assigned white, and the pixels with intensity values lower than the threshold are assigned a dark value. This is known as binary thresholding, and is the most basic form of thresholding and segmentation. An example is shown in Listing 9-9.

Listing 9-9. prog09.py

```python
import matplotlib.pyplot as plt
import scipy.misc as misc
img = misc.ascent()
thresh = img > 127
output = [img, thresh]
titles = ['Original', 'Thresholding']
for i in range(2):
        plt.subplot(1, 2, i+1)
        plt.imshow(output[i], cmap='gray')
        plt.title(titles[i])
        plt.axis('off')
plt.show()
```

The code in Listing 9-9 sets the threshold at 127. At the grayscale, a pixel value of 127 corresponds to the gray color. The resulting thresholded image is shown in Figure 9-9.

Original

Thresholding

Figure 9-9. *Binary thresholding*

Thresholding is the most basic type of image segmentation. Image segmentation refers to dividing an image into many regions based on some property, like colors of pixels, connectivity of the region, and so forth. You can get the better segments of an image by applying morphological operations to a thresholded image (see Listing 9-10).

Listing 9-10. prog10.py

```python
import matplotlib.pyplot as plt
import scipy.misc as misc
import scipy.ndimage as ndi
import numpy as np
img = misc.ascent()
thresh = img > 127
dilated = ndi.binary_dilation(thresh, structure=np.ones((9,
9))).astype(int)
eroded = ndi.binary_erosion(dilated, structure=np.ones((9,
9))).astype(int)
output = [img, thresh, dilated, eroded]
```

```
titles = ['Original', 'Thresholding', 'Dilated', 'Eroded and
Segmented']
for i in range(4):
    plt.subplot(2, 2, i+1)
    plt.imshow(output[i], cmap='gray')
    plt.title(titles[i])
    plt.axis('off')
plt.show()
```

The output is shown in Figure 9-10.

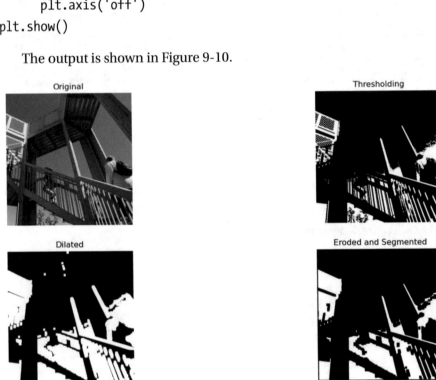

Figure 9-10.   *Thresholded and segmented image*

The top-hat transformation extracts small elements from image. we can compute white tophat and black tophat as demonstrated in Listing 9-11.

Listing 9-11. prog11.py

```python
import matplotlib.pyplot as plt
import scipy.misc as misc
import scipy.ndimage as ndi
img = misc.ascent()
struct = ndi.generate_binary_structure(rank=2,
                                        connectivity=3)
o1 = ndi.black_tophat(input=img,
                      structure=struct)
o2 = ndi.white_tophat(input=img,
                      structure=struct)
output = [img, o1, o2]
titles = ['Original', 'Black Tophat',
          'White Tophat']
for i in range(3):
        plt.subplot(1, 3, i+1)
        plt.imshow(output[i],
                   cmap='gray')
        plt.title(titles[i])
        plt.axis('off')
plt.show()
```

The output is shown in Figure 9-11.

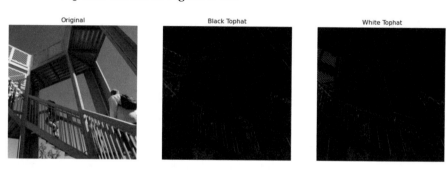

Figure 9-11. *Black and white tophat*

185

We can also iterate over a structure, as demonstrated in Listing 9-12.

Listing 9-12. prog12.py

```
import scipy.ndimage as ndi
struct = ndi.generate_binary_structure(2, 1).astype(int)
print(struct)
print(ndi.iterate_structure(struct, 2).astype(int))
print(ndi.iterate_structure(struct, 3).astype(int))
```

The output is as follows:

```
[[0 1 0]
 [1 1 1]
 [0 1 0]]

[[0 0 1 0 0]
 [0 1 1 1 0]
 [1 1 1 1 1]
 [0 1 1 1 0]
 [0 0 1 0 0]]

[[0 0 0 1 0 0 0]
 [0 0 1 1 1 0 0]
 [0 1 1 1 1 1 0]
 [1 1 1 1 1 1 1]
 [0 1 1 1 1 1 0]
 [0 0 1 1 1 0 0]
 [0 0 0 1 0 0 0]]
```

We can also compute the morphological gradient, as demonstrated in Listing 9-13.

Listing 9-13. prog13.py

```python
import matplotlib.pyplot as plt
import scipy.misc as misc
import scipy.ndimage as ndi
img = misc.ascent()
o1 = ndi.morphological_gradient(img, size=(3, 3))
struct = ndi.generate_binary_structure(2, 1)
o2 = ndi.morphological_laplace(img,
                                structure=struct)
output = [img, o1, o2]
titles = ['Original',
          'M Gradient',
          'M Laplace']
for i in range(3):
        plt.subplot(1, 3, i+1)
        plt.imshow(output[i],
                  cmap='gray')
        plt.title(titles[i])
        plt.axis('off')
plt.show()
```

The output is shown in Figure 9-12.

  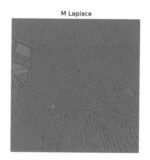

Figure 9-12. *Morphological gradients*

Summary

In this chapter, we studied distance transforms for generating test images. We then learned about morphology and how to use morphological operations on images. Morphological operations come in two varieties— binary and grayscale. We also studied thresholding, which is the simplest form of image segmentation.

In the next chapter, we will study and demonstrate video processing.

CHAPTER 10

Video Processing

In the previous chapter, we learned about using **morphological operations** in the domain of image processing. We have also demonstrated all those operations with libraries such as NumPy, SciPy, and matplotlib.

Till now, throughout this book, we have been discussing how to work with static images. We know the means of acquiring static images and also how to process them. This chapter is dedicated to processing streams of continuous images, such as video and live webcam feed. We will also get introduced to a new library, OpenCV. We will cover the following topics in this chapter:

- Introduction to OpenCV library

- Converting the colorspace of image

- Separating color channels in a live webcam stream

- More operations on live webcam streams

By the end of this chapter, we will be very comfortable with processing webcam feeds and videos with SciPy and NumPy.

Introduction to OpenCV Library

We have explored how many functions found in the SciPy and NumPy libraries can be used to process images. We will now examine a very limited part of another popular image processing library, OpenCV. We

© Ashwin Pajankar 2022
A. Pajankar, *Raspberry Pi Image Processing Programming*,
https://doi.org/10.1007/978-1-4842-8270-0_10

are going to use it for reading live webcam stream and video files. For processing the video and live webcam feed, we will still use SciPy.

OpenCV stands for **open source computer vision**. It is a library for real-time computer vision. It includes a lot of functionality for image processing and video procession. It also has implementations of many computer vision algorithms. It is primarily written in C++. It has interfaces (APIs; application programming interfaces) for languages such as C++, Python, Java, MATLAB/OCTAVE, and JavaScript. We can explore this at the webpage https://opencv.org/.

Let's get started by installing the library on our Raspberry Pi with the following command:

```
pip install opencv-python
```

For our further demonstrations, we will be needing a webcam. I have been using a **Logitech C922 Pro**. The camera is quite costly, with a price around 150 USD. However, I use it for multiple applications. If this one is out of your budget, you can use a less expensive webcam for the demonstrations. Most of the branded USB webcams will just do fine with Raspberry Pi. You can also check the list of webcams compatible with RPi at the webpage https://elinux.org/RPi_USB_Webcams. To get started, connect the webcam to the Raspberry Pi using one of its four USB ports.

Now, have a look at the following code listing (Listing 10-1).

Listing 10-1. prog00.py

```
import numpy as np
import cv2 as cv

cap = cv.VideoCapture(0)

cap.set(cv.CAP_PROP_FRAME_WIDTH, 256)
cap.set(cv.CAP_PROP_FRAME_HEIGHT, 144)
```

```
if not cap.isOpened():
    print("Cannot open camera...")
    exit()

while True:
    ret, frame = cap.read()

    if not ret:
        print("Can't receive frame... \nExiting ...")
        break

    cv.imshow('frame', frame)
    if cv.waitKey(1) == 27:
        break

cap.release()
cv.destroyAllWindows()
```

Execute the program. The indicator LED embedded in the webcam will glow, and it will produce the following (Figure 10-1) output window on the display,

Figure 10-1. *The output of the live webcam feed*

That's me! We can stop the program by pressing the ESC key on the keyboard. The window (Figure 10-1) will close, and the LED indicator in the webcam will stop glowing. Now, it is time to understand the program line-by-line. In the first two lines, we import the required libraries (which includes OpenCV) and create aliases for our convenience. Then, the following line creates an object corresponding to the webcam:

```
cap = cv.VideoCapture(0)
```

The argument passed is the index of the webcam. If only one webcam is attached then the argument passed is always 0. The following two lines define the resolution at which the webcam will operate:

```
cap.set(cv.CAP_PROP_FRAME_WIDTH, 256)
cap.set(cv.CAP_PROP_FRAME_HEIGHT, 144)
```

The function cap.set() sets the properties of the webcam connected to the computer. There are plenty of properties. The following (https://stackoverflow.com) question discusses the properties of webcams that can be set with OpenCV in detail:

https://stackoverflow.com/questions/11420748/setting-camera-parameters-in-opencv-python

Also, I wish to make readers aware that we will find the video stream from the webcam faster and more responsive at lower resolutions. As the processing power of the Raspberry Pi is limited, we will have more **frames per second** (FPS) in the output window at lower resolutions. We can get a list of all the standard resolutions at the following resources on the web:

https://en.wikipedia.org/wiki/List_of_common_resolutions
https://yusef.es/labs/resolutions.htm

The following lines of code terminate the execution of the program if the webcam cannot be accessed:

```
if not cap.isOpened():
    print("Cannot open camera...")
    exit()
```

The while loop block runs as long as we do not hit the ESC key on the keyboard.

```
while True:
    ret, frame = cap.read()
```

The function call `cap.read()` returns the status and the frame. The frame is a numpy Ndarray object that has the current frame. We can manipulate it as usual with NumPy and SciPy routines, just as we have practiced in earlier chapters.

If the status of the webcam frame read operation is unsuccessful, then we terminate the loop, as follows:

```
if not ret:
    print("Can't receive frame... \nExiting ...")
    break
```

The following routine call shows the frame:

```
cv.imshow('frame', frame)
```

The following code checks for the press of the ESC key on the keyboard:

```
if cv.waitKey(1) == 27:
    break
```

The following code releases the webcam by destroying the object:

```
cap.release()
```

The following code destroys all the current OpenCV windows:

```
cv.destroyAllWindows()
```

Consider this program as the template for all the programs in the rest of the chapter. We can modify the program to read and show a video file. We have to modify the code where we address the webcam, as follows:

```
cap = cv.VideoCapture('test.mp4')
```

We are passing the name of the video file as the argument. We also have to make modifications to the following line in the while loop block:

```
    if cv.waitKey(1) == 27:
```

The argument we are passing while using the webcam is 1. However, to show the frames of the video file, we have to compute the value of this argument we are passing to this function call. We first need to know the frame rate of the video in terms of FPS and then divide 1000 by that rate. Then, we have to pass the output of this computation as the argument to the function call cv.waitKey(). Let us see a few examples with standard frame rates. Usually, the cinematic frame rate is around 24 FPS (23.97, actually). So, 1000/24 = 41.67, which is almost equal to 42. So, the function call for 24 FPS will be cv.waitKey(42). For 25 FPS, it is cv.waitKey(40). For 30 FPS, it is cv.waitKey(33), and so on. As an exercise, open a video file by making the discussed modifications.

Converting the Colorspace of Image

Colorspace is a mathematical representation of color models. A color model is a way to represent the colors in a reproducible way. The OpenCV library reads an image or a live webcam feed or a video in the BGR colorspace. We can convert an image from BGR to any colorspace of our choice. We will frequently need to convert the live webcam feed from BGR to grayscale throughout the chapter. Let's see how to do that. Have a look at Listing 10-2.

Listing 10-2. prog01.py

```python
import numpy as np
import cv2 as cv

cap = cv.VideoCapture(0)
cap.set(cv.CAP_PROP_FRAME_WIDTH, 256)
cap.set(cv.CAP_PROP_FRAME_HEIGHT, 144)

if not cap.isOpened():
    print("Cannot open camera...")
    exit()

while True:
    ret, frame = cap.read()

    if not ret:
        print("Can't receive frame... \nExiting ...")
        break

    gray = cv.cvtColor(frame,
                       cv.COLOR_BGR2GRAY)

    cv.imshow('frame', gray)
    if cv.waitKey(1) == 27:
        break

cap.release()
cv.destroyAllWindows()
```

In the following line, we are converting the color frame into a greyscale frame,

```python
    gray = cv.cvtColor(frame,
                       cv.COLOR_BGR2GRAY)
```

The function cvtColor() is used to achieve the conversion between colorspaces. We will also be using this function quite frequently throughout the chapter as many image processing routines in SciPy need grayscale images as the input. Run the program to see the output.

Separating Color Channels in Live Webcam Stream

We can also separate the color channels in the live webcam stream with the code in Listing 10-3.

Listing 10-3. prog02.py

```python
import numpy as np
import cv2 as cv
from datetime import datetime

cap = cv.VideoCapture(0)

cap.set(cv.CAP_PROP_FRAME_WIDTH, 256)
cap.set(cv.CAP_PROP_FRAME_HEIGHT, 144)

if not cap.isOpened():
    print("Cannot open camera...")
    exit()

while True:
    ret, frame = cap.read()

    if not ret:
        print("Can't receive frame... \nExiting ...")
        break

    print(datetime.now())
```

```
print(frame.dtype)
print(frame.shape)
print(frame.ndim)
print(frame.size)

b = frame[:, :, 0]
g = frame[:, :, 1]
r = frame[:, :, 2]

cv.imshow('Red', r)
cv.imshow('Green', g)
cv.imshow('Blue', b)
if cv.waitKey(1) == 27:
    break

cap.release()
cv.destroyAllWindows()
```

In this code listing (prog02.py), we can see the properties of each frame as well as all the channels, as shown in the output (Figure 10-2).

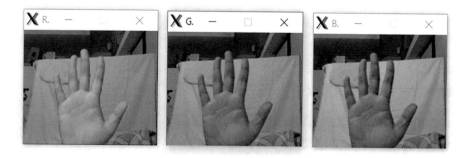

Figure 10-2. *Seperation of color channels*

Also, the console will show the output, like the one here, several times per second (so, it will scroll really fast):

```
2022-01-30 13:00:30.427360
```

```
uint8
(144, 176, 3)
3
76032
```

More Operations on Live Webcam Stream

Let us see more operations we can perform on a live webcam stream with the SciPy library routines. We are going to cover a lot of examples. We learned all these operations in earlier chapters, and we will now apply them on a live webcam stream.

Geometric Transformation

Let's rotate the webcam frame continuously. Listing 10-4 shows the code.

Listing 10-4. prog03.py

```python
import numpy as np
import cv2 as cv
import scipy.ndimage as ndi

angle = 0
cap = cv.VideoCapture(0)

cap.set(cv.CAP_PROP_FRAME_WIDTH, 256)
cap.set(cv.CAP_PROP_FRAME_HEIGHT, 144)

if not cap.isOpened():
    print("Cannot open camera...")
    exit()

while True:
    ret, frame = cap.read()
```

```
if not ret:
    print("Can't receive frame... \nExiting ...")
    break

angle = (angle + 1) % 360
output = ndi.rotate(frame, angle,
                    reshape=False)

cv.imshow('Rotated Frame', output)
if cv.waitKey(1) == 27:
    break
cap.release()
cv.destroyAllWindows()
```

The code that manipulates the frame is the following one:

```
angle = (angle + 1) % 360
output = ndi.rotate(frame, angle,
                    reshape=False)
```

The logic is surprisingly plain. We just increment the rotation angle with each frame, and when it reaches 360 degrees, we start over. The following image (Figure 10-3) shows a screenshot of one of the frames.

Figure 10-3. *Geometric transformation on a live webcam feed*

Actually, the screenshot (Figure 10-3) really does not do any justice to the real output. Run the program and see the output for yourself.

Convolution

Let's define a kernel and apply it to the frames in the stream using the convolution operation, as in Listing 10-5.

Listing 10-5. prog04.py

```python
import numpy as np
import cv2 as cv
import scipy.ndimage as ndi

cap = cv.VideoCapture(0)
n = 5
k = np.ones((n, n, n), np.uint8)/(n*n*n)

cap.set(cv.CAP_PROP_FRAME_WIDTH, 256)
cap.set(cv.CAP_PROP_FRAME_HEIGHT, 144)

if not cap.isOpened():
    print("Cannot open camera...")
    exit()

while True:
    ret, frame = cap.read()

    if not ret:
        print("Can't receive frame... \nExiting ...")
        break

    output = ndi.convolve(frame, k,
                          mode='nearest')

    cv.imshow('Convolved Frame', output)
```

```
    if cv.waitKey(1) == 27:
        break
cap.release()
cv.destroyAllWindows()
```

Since it is a blurring kernel, the output is a blurred feed from the webcam, as shown in the screenshot (Figure 10-4).

Figure 10-4. *Blurred feed from the webcam*

The OpenCV library also has the functionality to create a trackbar. Let's create a trackbar in such a way that it adjusts the size of the kernel matrix. Listing 10-6 contains the code.

Listing 10-6. prog05.py

```
import numpy as np
import cv2 as cv
import scipy.ndimage as ndi

def nothing(x):
    pass

WINDOW_NAME = "Live Convolution"
cv.namedWindow(WINDOW_NAME)
cv.createTrackbar('Kernel Size',
```

```
                    WINDOW_NAME,
                    1, 5, nothing)
cap = cv.VideoCapture(0)
n = 1

cap.set(cv.CAP_PROP_FRAME_WIDTH, 256)
cap.set(cv.CAP_PROP_FRAME_HEIGHT, 144)

if not cap.isOpened():
    print("Cannot open camera...")
    exit()

while True:
    ret, frame = cap.read()

    if not ret:
        print("Can't receive frame... \nExiting ...")
        break

    n = cv.getTrackbarPos('Kernel Size',
                        WINDOW_NAME)
    k = np.ones((n, n, n), np.uint8)/(n*n*n)
    output = ndi.convolve(frame, k,
                        mode='nearest')

    cv.imshow(WINDOW_NAME, output)
    if cv.waitKey(1) == 27:
        break

cap.release()
cv.destroyAllWindows()
```

We are already familiar with most of the code. Let's understand the new parts. The following code defines an empty function:

```
def nothing(x):
    pass
```

The following code defines a named window:

```
WINDOW_NAME = "Live Convolution"
cv.namedWindow(WINDOW_NAME)
```

We need a named window to bind the newly created trackbar as follows:

```
cv.createTrackbar('Kernel Size',
                  WINDOW_NAME,
                  1, 5, nothing)
```

This code creates a named trackbar that allows us to slide from 1 to 5. It is attached to the earlier-created named window. The empty function we defined earlier is associated with the trackbar and is called every time we operate the trackbar. We get the current position of the slider with the following function call in the while loop:

```
n = cv.getTrackbarPos('Kernel Size', WINDOW_NAME)
```

We can now change the size of the kernel by sliding the tracker on the trackbar. Figure 10-5 shows the output.

Figure 10-5. *Using trackbar*

We are going to use the code for the trackbar frequently throughout the chapter.

Correlation

Let's use the trackbar and apply a correlation on the live webcam feed, as shown in Listing 10-7.

Listing 10-7. prog06.py

```python
import numpy as np
import cv2 as cv
import scipy.ndimage as ndi

def nothing(x):
    pass

WINDOW_NAME = "Live Correlation"
cv.namedWindow(WINDOW_NAME)
cv.createTrackbar('Kernel Size',
                  WINDOW_NAME,
                  1, 5, nothing)
cap = cv.VideoCapture(0)
n = 1

cap.set(cv.CAP_PROP_FRAME_WIDTH, 256)
cap.set(cv.CAP_PROP_FRAME_HEIGHT, 144)

if not cap.isOpened():
    print("Cannot open camera...")
    exit()

while True:
    ret, frame = cap.read()
```

```
if not ret:
    print("Can't receive frame... \nExiting ...")
    break

n = cv.getTrackbarPos('Kernel Size',
                    WINDOW_NAME)
k = np.ones((n, n, n), np.uint8)/(n*n*n)
output = ndi.correlate(frame, k,
                    mode='nearest')

cv.imshow(WINDOW_NAME, output)
if cv.waitKey(1) == 27:
    break
cap.release()
cv.destroyAllWindows()
```

Let's see the output in Figure 10-6.

Figure 10-6. *Correlation with trackbar*

Note   We can read more about the concept of correlation at the URL https://towardsdatascience.com/convolution-vs-correlation-af868b6b4fb5.

Filtering

We have learned about many filters in earlier chapters. Let's apply them on the live webcam feed one by one. Listing 10-8 shows the Gaussian filter.

Listing 10-8. prog07.py

```python
import numpy as np
import cv2 as cv
import scipy.ndimage as ndi

def nothing(x):
    pass

WINDOW_NAME = "Live Gaussian"
cv.namedWindow(WINDOW_NAME)
cv.createTrackbar('Sigma Size',
                WINDOW_NAME,
                1, 15, nothing)
cap = cv.VideoCapture(0)

cap.set(cv.CAP_PROP_FRAME_WIDTH, 256)
cap.set(cv.CAP_PROP_FRAME_HEIGHT, 144)

if not cap.isOpened():
    print("Cannot open camera...")
    exit()

while True:
```

```
    ret, frame = cap.read()

    if not ret:
        print("Can't receive frame... \nExiting ...")
        break

    n = cv.getTrackbarPos('Sigma Size',
                          WINDOW_NAME)

    output = ndi.gaussian_filter(frame,
                                 sigma=n)

    cv.imshow(WINDOW_NAME, output)
    if cv.waitKey(1) == 27:
        break

cap.release()
cv.destroyAllWindows()
```

Let's apply a uniform filter, as shown in Listing 10-9.

Listing 10-9. prog08.py

```
import numpy as np
import cv2 as cv
import scipy.ndimage as ndi

def nothing(x):
    pass

WINDOW_NAME = "Live Uniform"
cv.namedWindow(WINDOW_NAME)
cv.createTrackbar('Filter Size',
                  WINDOW_NAME,
                  1, 15, nothing)
cap = cv.VideoCapture(0)
```

```python
cap.set(cv.CAP_PROP_FRAME_WIDTH, 256)
cap.set(cv.CAP_PROP_FRAME_HEIGHT, 144)

if not cap.isOpened():
    print("Cannot open camera...")
    exit()

while True:
    ret, frame = cap.read()

    if not ret:
        print("Can't receive frame... \nExiting ...")
        break

    n = cv.getTrackbarPos('Filter Size',
                          WINDOW_NAME)

    output = ndi.uniform_filter(frame,
                               size=n)

    cv.imshow(WINDOW_NAME, output)
    if cv.waitKey(1) == 27:
        break

cap.release()
cv.destroyAllWindows()
```

Let's apply the percentile filter. The code in Listing 10-10 defines two trackbars for adjusting the arguments passed to the filter.

Listing 10-10. prog09.py

```python
import numpy as np
import cv2 as cv
import scipy.ndimage as ndi

def nothing(x):
    pass
```

```python
WINDOW_NAME = "Live Percentile"
cv.namedWindow(WINDOW_NAME)
cv.createTrackbar('Size',
                  WINDOW_NAME,
                  1, 5, nothing)
cv.createTrackbar('Percentile',
                  WINDOW_NAME,
                  1, 75, nothing)
cap = cv.VideoCapture(0)

cap.set(cv.CAP_PROP_FRAME_WIDTH, 256)
cap.set(cv.CAP_PROP_FRAME_HEIGHT, 144)

if not cap.isOpened():
    print("Cannot open camera...")
    exit()

while True:
    ret, frame = cap.read()

    if not ret:
        print("Can't receive frame... \nExiting ...")
        break

    s = cv.getTrackbarPos('Size',
                          WINDOW_NAME)

    p = cv.getTrackbarPos('Percentile',
                          WINDOW_NAME)

    output = ndi.percentile_filter(frame, size=s,
                                   percentile=p)

    cv.imshow(WINDOW_NAME, output)
    if cv.waitKey(1) == 27:
        break
```

```
cap.release()
cv.destroyAllWindows()
```

The output is as shown in Figure 10-7.

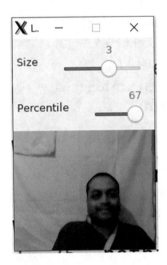

Figure 10-7. *Percentile filter*

Let's implement a minimum filter as shown in Listing 10-11.

Listing 10-11. prog10.py

```
import numpy as np
import cv2 as cv
import scipy.ndimage as ndi

def nothing(x):
    pass

WINDOW_NAME = "Live Minimum"
cv.namedWindow(WINDOW_NAME)
cv.createTrackbar('Filter Size',
                  WINDOW_NAME,
                  1, 55, nothing)
```

```python
cap = cv.VideoCapture(0)

cap.set(cv.CAP_PROP_FRAME_WIDTH, 256)
cap.set(cv.CAP_PROP_FRAME_HEIGHT, 144)

if not cap.isOpened():
    print("Cannot open camera...")
    exit()

while True:
    ret, frame = cap.read()

    if not ret:
        print("Can't receive frame... \nExiting ...")
        break

    n = cv.getTrackbarPos('Filter Size',
                        WINDOW_NAME)

    output = ndi.minimum_filter(frame,
                                size=n)

    cv.imshow(WINDOW_NAME, output)
    if cv.waitKey(1) == 27:
        break

cap.release()
cv.destroyAllWindows()
```

Let's apply the maximum filter, as in Listing 10-12.

Listing 10-12. prog11.py

```python
import numpy as np
import cv2 as cv
import scipy.ndimage as ndi
```

```python
def nothing(x):
    pass

WINDOW_NAME = "Live Maximum"
cv.namedWindow(WINDOW_NAME)
cv.createTrackbar('Filter Size',
                    WINDOW_NAME,
                    1, 25, nothing)
cap = cv.VideoCapture(0)

cap.set(cv.CAP_PROP_FRAME_WIDTH, 256)
cap.set(cv.CAP_PROP_FRAME_HEIGHT, 144)

if not cap.isOpened():
    print("Cannot open camera...")
    exit()

while True:
    ret, frame = cap.read()

    if not ret:
        print("Can't receive frame... \nExiting ...")
        break

    n = cv.getTrackbarPos('Filter Size',
                            WINDOW_NAME)

    output = ndi.maximum_filter(frame,
                                    size=n)

    cv.imshow(WINDOW_NAME, output)
    if cv.waitKey(1) == 27:
        break

cap.release()
cv.destroyAllWindows()
```

We can also apply Prewitt and Laplace filters as shown in Listing 10-13.

Listing 10-13. prog12.py

```python
import numpy as np
import cv2 as cv
import scipy.ndimage as ndi

cap = cv.VideoCapture(0)

cap.set(cv.CAP_PROP_FRAME_WIDTH, 256)
cap.set(cv.CAP_PROP_FRAME_HEIGHT, 144)

if not cap.isOpened():
    print("Cannot open camera...")
    exit()

while True:
    ret, frame = cap.read()

    if not ret:
        print("Can't receive frame... \nExiting ...")
        break

    o1 = ndi.prewitt(frame)
    o2 = ndi.laplace(frame)

    cv.imshow("Prewitt", o1)
    cv.imshow("Laplace", o2)
    if cv.waitKey(1) == 27:
        break

cap.release()
cv.destroyAllWindows()
```

Note   We can read more about both filters at the following URLs:

https://www.geeksforgeeks.org/matlab-image-edge-detection-using-prewitt-operator-from-scratch/

https://www.l3harrisgeospatial.com/docs/laplacianfilters.html

Let's apply the Gaussian gradient magnitude filter as in Listing 10-14.

Listing 10-14. prog13.py

```python
import numpy as np
import cv2 as cv
import scipy.ndimage as ndi

def nothing(x):
    pass

WINDOW_NAME = "Gaussian Gradient Magnitude"
cv.namedWindow(WINDOW_NAME)
cv.createTrackbar('Sigma',
                WINDOW_NAME,
                1, 10, nothing)
cap = cv.VideoCapture(0)

cap.set(cv.CAP_PROP_FRAME_WIDTH, 256)
cap.set(cv.CAP_PROP_FRAME_HEIGHT, 144)

if not cap.isOpened():
    print("Cannot open camera...")
    exit()

while True:
    ret, frame = cap.read()
```

```
    if not ret:
        print("Can't receive frame... \nExiting ...")
        break

    s = cv.getTrackbarPos('Sigma',
                          WINDOW_NAME)

    gray = cv.cvtColor(frame, cv.COLOR_BGR2GRAY)

    output = ndi.gaussian_gradient_magnitude(gray,
                                             sigma=s)

    cv.imshow(WINDOW_NAME, output)
    if cv.waitKey(1) == 27:
        break
cap.release()
cv.destroyAllWindows()
```

Let's apply the Gaussian Laplace filter as in Listing 10-15.

Listing 10-15. prog14.py

```
import numpy as np
import cv2 as cv
import scipy.ndimage as ndi

def nothing(x):
    pass

WINDOW_NAME = "Gaussian Laplace"
cv.namedWindow(WINDOW_NAME)
cv.createTrackbar('Sigma',
                 WINDOW_NAME,
                 1, 10, nothing)
cap = cv.VideoCapture(0)
```

```python
cap.set(cv.CAP_PROP_FRAME_WIDTH, 256)
cap.set(cv.CAP_PROP_FRAME_HEIGHT, 144)

if not cap.isOpened():
    print("Cannot open camera...")
    exit()

while True:
    ret, frame = cap.read()

    if not ret:
        print("Can't receive frame... \nExiting ...")
        break

    s = cv.getTrackbarPos('Sigma',
                          WINDOW_NAME)

    gray = cv.cvtColor(frame, cv.COLOR_BGR2GRAY)

    output = ndi.gaussian_laplace(gray,
                                  sigma=s)

    cv.imshow(WINDOW_NAME, output)
    if cv.waitKey(1) == 27:
        break

cap.release()
cv.destroyAllWindows()
```

Morphological Operations

We can also apply morphological operations on live video, as shown in Listing 10-16.

Listing 10-16. prog15.py

```python
import numpy as np
import cv2 as cv
import scipy.ndimage as ndi

def nothing(x):
    pass

WINDOW_NAME = "Distance Transform"
cv.namedWindow(WINDOW_NAME)

cv.createTrackbar('Sigma',
                  WINDOW_NAME,
                  1, 10, nothing)

cap = cv.VideoCapture(0)

cap.set(cv.CAP_PROP_FRAME_WIDTH, 256)
cap.set(cv.CAP_PROP_FRAME_HEIGHT, 144)

if not cap.isOpened():
    print("Cannot open camera...")
    exit()

while True:
    ret, frame = cap.read()

    if not ret:
        print("Can't receive frame... \nExiting ...")
        break

    grey = cv.cvtColor(frame, cv.COLOR_BGR2GRAY)

    n1 = cv.getTrackbarPos('Sigma',
                           WINDOW_NAME)
```

```
output = ndi.grey_dilation(grey, size=(7, 7),
                            structure=np.ones((n1, n1)))

    cv.imshow(WINDOW_NAME, output)
    if cv.waitKey(1) == 27:
        break

cap.release()
cv.destroyAllWindows()
```

This code applies the grayscale dilation to a live webcam stream, as shown in Figure 10-8.

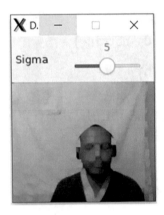

Figure 10-8. *Grayscale dilation*

We can similarly apply erosion to a live feed, as shown in Listing 10-17.

Listing 10-17. prog16.py

```
import numpy as np
import cv2 as cv
import scipy.ndimage as ndi

def nothing(x):
    pass
```

```python
WINDOW_NAME = "Distance Transform"
cv.namedWindow(WINDOW_NAME)

cv.createTrackbar('Sigma',
                  WINDOW_NAME,
                  1, 10, nothing)

cap = cv.VideoCapture(0)

cap.set(cv.CAP_PROP_FRAME_WIDTH, 256)
cap.set(cv.CAP_PROP_FRAME_HEIGHT, 144)

if not cap.isOpened():
    print("Cannot open camera...")
    exit()

while True:
    ret, frame = cap.read()

    if not ret:
        print("Can't receive frame... \nExiting ...")
        break

    grey = cv.cvtColor(frame, cv.COLOR_BGR2GRAY)

    n1 = cv.getTrackbarPos('Sigma',
                           WINDOW_NAME)

    output = ndi.grey_erosion(grey, size=(7, 7),
                              structure=np.ones((n1, n1)))

    cv.imshow(WINDOW_NAME, output)
    if cv.waitKey(1) == 27:
        break
```

```
cap.release()
cv.destroyAllWindows()
```

The output is as seen in Figure 10-9.

Figure 10-9. *Grayscale erosion*

Summary

In this chapter, we learned how to apply the SciPy routines to a live video feed from a webcam and on video files. We are now comfortable with processing videos (both live and stored files).

Conclusion

In this book, we covered various Python libraries meant for image processing. We also used the Raspberry Pi 4 single-board computer with the Raspberry Pi OS Debian Buster as the preferred platform for demonstrating all the Python 3 code. All the code examples can be run on any other operating system as well, such as Windows, macOS, other Linux distributions, Unix distributions, and BSD (such as FreeBSD).

The world of image processing and computer vision is very vast. If we consider it as an ocean, then we are still on the shores of it. Equipped with basic knowledge, we can definitely venture ourselves into deeper waters and gain understanding of computer vision in greater depth. Armed with the advanced knowledge, we can program various real-life examples for computer vision.

Appendix

This section examines topics that will help us to work with image processing and Raspberry Pi SBC in a more efficient way. While I could not find a suitable place for these topics in any of the chapters, I personally feel that they are quite useful in the learning process. This appendix section covers the following topics:

- pgmagick image-processing library

- Connecting a display

- Connecting to ethernet/wired network

- Remote desktop with Virtual Network Computing (VNC)

After reviewing the appendix, you should be comfortable with these tips and tricks for Raspberry Pi.

pgmagick Image Processing

Here we will work with a simple image-processing library known as **pgmagick**. It is a Python wrapper around the image-manipulation program **ImageMagik**. One can read more about the ImageMagik program at its homepage: https://imagemagick.org/index.php.

Let's install everything by running the following commands in sequence:

```
sudo apt-get install imagemagick libgraphicsmagick++1-dev -y
sudo apt-get install g++ libboost-python-dev -y
sudo pip3 install pgmagick
```

© Ashwin Pajankar 2022
A. Pajankar, *Raspberry Pi Image Processing Programming*,
https://doi.org/10.1007/978-1-4842-8270-0

It installs everything required to write a Python program using the pgmagick library. Listing A-1 creates a simple background with green color and saves that as an image in the current directory.

Listing A-1. prog00.py

```
from pgmagick.api import Image
img = Image((300, 200), 'Green')
img.write('bg-green.png')
```

Let's create a transparent background as in Listing A-2.

Listing A-2. prog01.py

```
from pgmagick.api import Image
img = Image((300, 200), 'transparent')
img.write('transparent.png')
```

Let's create a gradient image as shown in Listing A-3.

Listing A-3. prog02.py

```
from pgmagick.api import Image
img = Image((300, 200),
            'gradient:#00ff00-#0000ff')
img.write('gradient.png')
```

We can add some text to an image with the code shown in Listing A-4.

Listing A-4. prog03.py

```
from pgmagick.api import Image
img = Image((300, 200))
img.annotate('Hello World', angle=60)
img.write('helloworld60.png')
```

We can also retrieve the properties of an image as shown in Listing A-5.

Listing A-5. prog04.py

```
from pgmagick.api import Image
img = Image((300, 200))
print (img.width, img.height)
```

These are the basics of the library. Please run all the preceding programs and see the output. One can explore this library further at https://pythonhosted.org/pgmagick/.

Connecting a Display

Up until now, we have been accessing the Raspberry Pi in headless mode (connecting using SSH and X-11 forwarding using Wi-Fi). We can also connect a visual display (an HDMI/VGA monitor) by choosing the appropriate connector. If we are opting for HDMI with RPi 4B, then we need a **micro-HDMI to HDMI** adapter (check online marketplaces for this). If we are opting for a VGA monitor, we need a **micro-HDMI to VGA** adapter. If we check online marketplaces, we can find a **micro-HDMI to HDMI and VGA combined adapter** that takes care of both cases.

Using a VGA Display

Additionally, if we wish to use a VGA display, then we must make changes to the /boot/config.txt file before booting up the first time. This step is needed only for a VGA monitor. If you are using an HDMI monitor, then skip this part.

Insert the microSD card into the card reader and browse it in **Windows Explorer** (or the file explorer of your operating system). In Windows Explorer, it will be represented as a removable media drive labeled as boot.

Open the `config.txt` file and make the following changes to it:

- Change #`disable_overscan=1` to `disable_overscan=1`

- Change #`hdmi_force_hotplug=1` to `hdmi_force_hotplug=1`

- Change #`hdmi_group=1` to `hdmi_group=2`

- Change #`hdmi_mode=1` to `hdmi_mode=16`

- Change #`hdmi_drive=2` to `hdmi_drive=2`

- Change #`config_hdmi_boost=4` to `config_hdmi_boost=4`

Save the file after making these changes. The microSD card is now ready for the Pi with a VGA monitor.

Booting Up After Connecting a Display

As we have seen, we can connect the Pi to the monitor of our choice and boot up. After booting up for the very first time, we will be greeted by a setup or configuration wizard, as shown in Figure A-1. It is a one-time process only, so go ahead.

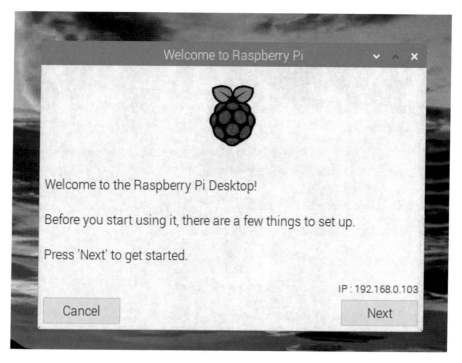

Figure A-1. *Configuration wizard*

Click on the **Next** button, and you will be shown the following window (Figure A-2).

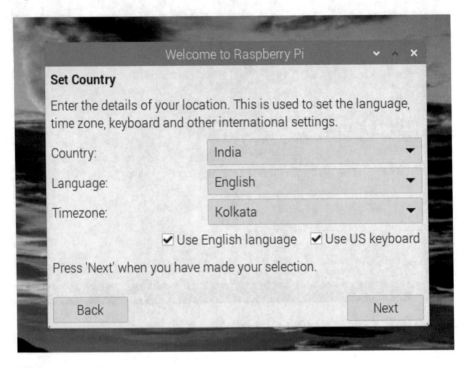

Figure A-2. Localization settings

Choose the most appropriate options for your region and click **Next** (you can also navigate to the previous options by clicking on **Back**). You will see the following (Figure A-3) window.

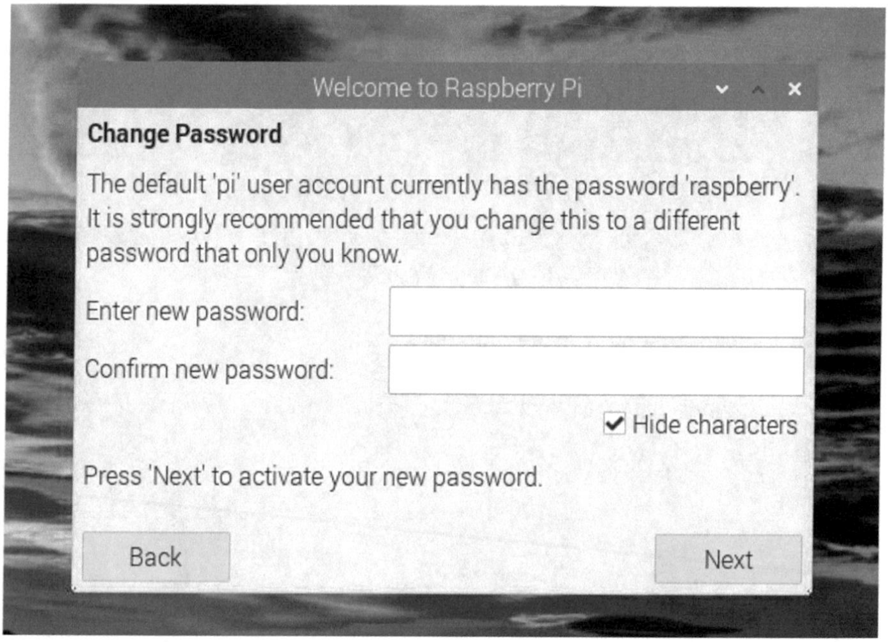

Figure A-3. *Changing the default password*

Change the default password. It is strongly recommended, as anyone who hacks into your Wi-Fi can remotely log in to your Pi using the default password and wreak havoc. Press the **Next** button, and it will show the following (Figure A-4) window.

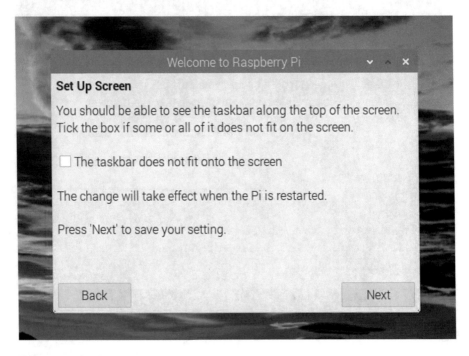

Figure A-4. *Set up screen*

Check the box if the taskbar (in the top area of the screen) does not fit the screen, and then click on **Next**. It shows the following (Figure A-5) window.

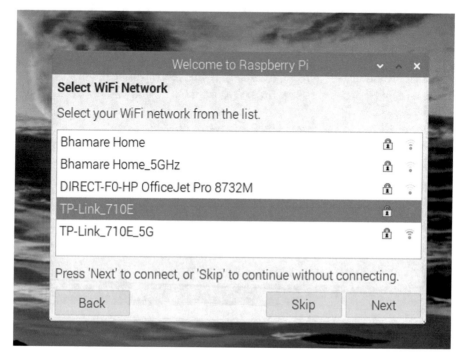

Figure A-5. *Set up the Wi-Fi*

We can set up the Wi-Fi here (if not set up already; if it is, click on the **Skip** button), and it shows the update window (Figure A-6).

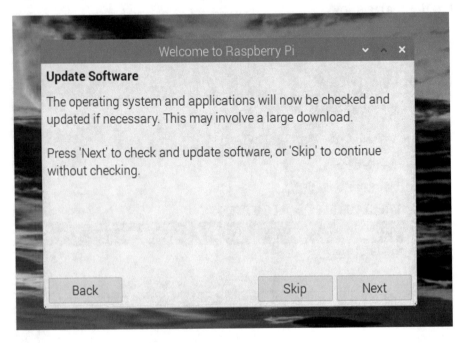

Figure A-6. *Software update*

We can click on the **Next** button, and it will update the software (we can also skip this, but I recommend updating the software). Once done, it will show a success message (Figure A-7).

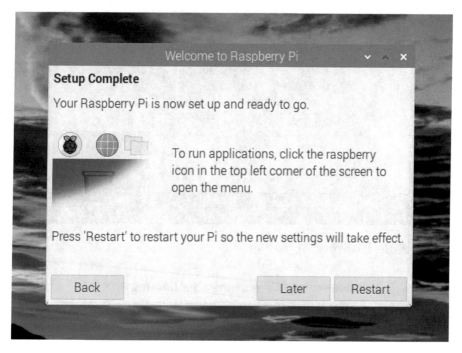

Figure A-7. *Success!*

Here, click on the **Restart** button. After reboot, if you have skipped changing the password, it will show the following warning (Figure A-8).

Figure A-8. *Warning*

We will continue seeing this warning after booting up as long as the password is the default one. Once we change the password, it will stop appearing after bootup.

This finishes setup.

Connecting to Ethernet/ Wired Network

Most of us will use the Pi boards in an organization (workplace or academic institution) or at home. These places will always have wired LANs that are connected to the internet. We can connect the Pi to a LAN using the RJ45 connector. Pi boards have an RJ45 port for wired networking. Almost all modern routers come with DHCP (Dynamic Host Configuration Protocol), and an IP address will automatically be assigned to the Pi once connected to a wired network.

Remote Desktop with VNC

We can access the desktop remotely using the VNC software. We have to enable it in the Pi before using it. We saw the **raspi-config** tool in the very first chapter. We can enable VNC using that (in the command line). We can also find a GUI utility for the same task in the desktop environment. Click the Raspberry symbol in the top left. It is a Windows-like menu. In the **Preferences** section, we can find the option **Raspberry Pi Configuration**. Click on that, and it will open a new window. There, click on the third tab, labeled **Interfaces**. It shows the following (Figure A-9).

Figure A-9. Interfaces

Here, make sure that at least the options **SSH** and **VNC** are enabled. I need other options for my development and programming lessons, so I enabled them all.

After this, download a VNC viewer client from any of the
following URLs:

`https://www.realvnc.com/en/connect/download/viewer/`

`https://www.tightvnc.com/download.php`

`https://sourceforge.net/projects/tigervnc/files/`
`stable/1.12.0/`

RealVNC is one of the most popular VNC clients and the one I use.
I will explain the setup process for this. If you wish to use the other
clients, please check the relevant documentation and tutorials. We first
have to configure a connection by providing the IP address, as follows
(Figure A-10).

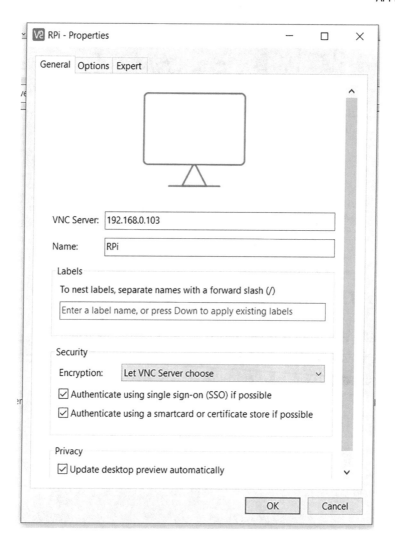

Figure A-10. *Configure the remote connection*

We also have to authenticate. I prefer to save the password so I won't have to enter it every time I try to connect (Figure A-11).

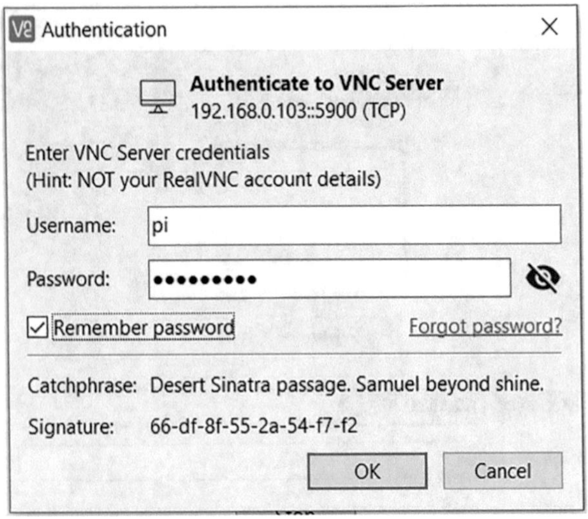

Figure A-11. *Enter and save the password*

Finally, if you are connecting for the very first time, it will show a warning message. Click on the **Always** button, and it will launch the remote desktop (Figure A-12).

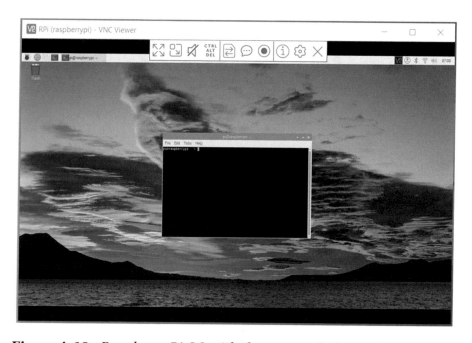

Figure A-12. *Raspberry Pi OS with the remote desktop using RealVNC*

Figure A-12 shows a remote desktop with a terminal client opened. In the top section, we can see the options for the VNC software. Explore them for better understanding.

Index

© Ashwin Pajankar 2022
A. Pajankar, *Raspberry Pi Image Processing Programming*,
https://doi.org/10.1007/978-1-4842-8270-0

Printed in the United States
by Baker & Taylor Publisher Services